# The Peripatetic University

*Unfinished Portrait of James Stuart in 1881 by Herkomer*
*(Reproduced by permission of Trinity College)*

# The Peripatetic University

## Cambridge Local Lectures
## 1873-1973

EDWIN WELCH

CAMBRIDGE
at the University Press
1973

CAMBRIDGE UNIVERSITY PRESS
Cambridge, New York, Melbourne, Madrid, Cape Town, Singapore, São Paulo, Delhi

Cambridge University Press
The Edinburgh Building, Cambridge CB2 8RU, UK

Published in the United States of America by Cambridge University Press, New York

www.cambridge.org
Information on this title: www.cambridge.org/9780521201520

First published 1973
This digitally printed version 2008

*A catalogue record for this publication is available from the British Library*

*Library of Congress Catalogue Card Number: 72-91961*

ISBN 978-0-521-20152-0 hardback
ISBN 978-0-521-08959-3 paperback

# Contents

# Preface

In 1967 Mr John Andrew, Secretary of the University of Cambridge's Board of Extra-Mural Studies, asked me to put in order the archives of the Board and its predecessor, the Local Lectures Syndicate. As the contents of the basement of Stuart House were sorted and catalogued, it became clear that all the papers had been kept – often in multiple copies – and that they formed a unique record of the origins and development of university extension. This was a whole new field of research on which little has been written and much of what has been written is inaccurate.* It was surprising to discover that, even within the University of Cambridge, it was not generally known that extra-mural studies were first established there, neither was it generally known how much pioneering work was undertaken by the Board.

When in October 1969 the Board invited me to write its history, I accepted with great interest. The Board gave me free access to all its archives and imposed no restrictions on their use. I have occasionally omitted details where living persons might still be hurt by their disclosure, but have added a reference to the original document. All expressions of opinion in this book are my own and not necessarily those of the Board. While writing I have been overwhelmed by the assistance which has been freely offered by everyone concerned. I was unable to use some of the information because of lack of time and lack of space. A list of those institutions and individuals from whom I have received help is given below, but I must here thank my wife who has patiently assisted me with the research and the writing.

In the space available I have tried to concentrate on certain themes – the early years when Cambridge stood alone, the continuity of the work from James Stuart's original plan, and the

* An exception must be made for Professor Kelly's work on the history of adult education which has proved invaluable in writing this book.

perpetual problem of finance. Some aspects of the Board's history have only just been touched upon, and it would have been possible to write a much longer book. Nevertheless I hope that this book conveys something of the importance and interest of its subject and that it repays to some extent the kindness of all who have assisted me.

E.W.

# Sources and Acknowledgements

The chief source of information is, of course, the Board's own archives which are cited in footnotes by the reference BEMS. Copies of the catalogue of the archives can be seen at the University Archives (where the originals will be deposited), the offices of the Historical Manuscripts Commission in London, and at Stuart House. All other sources would be readily identifiable from the footnotes and the list of acknowledgements below. The records of local organisations are usually held by the secretary *pro tem*.

*Bedford*. Public Library. University Extension Society.

*Bury St Edmunds*. Bury Free Press Ltd. County Record Office. Secretary of the Athenaeum.

*Cambridge*. Board of Extra-Mural Studies and its staff (past and present). Local Examinations Syndicate. University Archivist. Trinity and Newnham College Libraries. County Record Office. City Library.

*Colchester*. University Extension Society. Public Library.

*Derby*. University Extension Society. Public Library. W.E.A. Branch.

*Exeter*. University Registry. County Record Office. City Record Office. City Library.

*Hastings*. Public Library.

*Ipswich*. County Record Office. Public Library.

*Kettering*. Public Library.

*Leeds*. City Archives Dept. City Library.

*Leicester*. City Archives Dept. City Library. Vaughan College, University of Leicester.

*Liverpool*. University Archivist. City Record Office. City Library. Dept. of Extension Studies, University of Liverpool.

*London*. Public Record Office. Historical Manuscripts Commission. Guildhall Library. Greater London Record Office (County Hall). Dept. of Education and Science Library.

SOURCES AND ACKNOWLEDGEMENTS

*Lowestoft.* Literary and Scientific Association. Public Library.
*Maidstone.* County Record Office.
*Manchester.* University Library.
*Norwich.* City and County Record Office. City Library.
*Nottingham.* City Archives Dept. City Library.
*Ottawa.* Library of the Library School.
*Plymouth.* City Archives Dept. City Library.
*Portsmouth.* City Record Office.
*Sheffield.* Dept. of Archives and Local History, City Library.
The following individuals kindly allowed themselves to be interviewed, or gave information by letter:

Mr J. Aspinall of Lowestoft, Mr Humphrey Boardman of Norwich (James Stuart's nephew), Mr B. W. C. Green (former chief clerk of Board), Mr J. Grimwood-Taylor of Derby, Mr David Hardman, Mr G. F. Hickson, Mr F. Jacques, Secretary of the W.E.A. Eastern District, Alderman J. Lacey of Portsmouth and Sir James Matthews of Southampton (members of the first Portsmouth tutorial class), Miss A. R. Murray of Lowestoft, Miss A. Platt of Wellingborough, and Mrs Westbrook of the Local Centres Union.

Assistance was received from the following individuals:

Mr A. Chadwick of Burton on Trent, Miss M. S. Hodgson of the B.B.C. Written Archives Centre, Prof. T. Kelly of Liverpool University, Miss R. Pope of New Jersey and Prof. J. Roach of Sheffield.

# 1
# The Background

The origins of modern adult education can be traced back to the Evangelical Revival of the eighteenth century. Its purpose then was to enable men and women to read the Bible for themselves and there were considerable debates about the propriety of teaching writing or arithmetic, particularly if instruction was given on the Sabbath. Perhaps the earliest attempts of this kind were the Circulating Schools established in Wales by the Rev. Griffith Jones from 1737 onwards. The schoolmaster employed by him spent only a few months in each place teaching 'hired Servants, Day-labourers, and married Men and Women, as well as the younger Sort' how to read. Separate classes were held for adults in the evening in many places, and at Trelech an old man of 71 learnt to read and write. An unfortunate lawsuit brought the work to an end in 1780, but the Calvinistic Methodist minister Thomas Charles of Bala later revived the schools in North Wales.[1]

Slightly earlier another Welshman, Dr Thomas Bray, was encouraging the same kind of work in England. He established parochial libraries to make books available, and the S.P.C.K. (which he helped to found) provided charity schools to teach reading. An attempt in 1711 to establish evening classes for adults was not successful,[2] and in England adult education languished until the end of the eighteenth century. Although some Methodists learnt to read on conversion, there were no systematic attempts at providing adult education. John Wesley deplored 'the thought of our master's keeping an evening school. It would swallow up the time he ought to have for his own improvement.'[3] However in 1812, after Wesley's death, the Bristol Methodists helped to form a class to teach adults reading only.[4]

To reach those children who were unable to attend school on weekdays, Robert Raikes started the first Sunday School at

[1] References will be found on p. 196.

Gloucester in 1780. The movement spread through England rapidly, and it was not long before similar schools for adults were being planned.* The first adult school was founded at Nottingham in 1798 by William Singleton who was a member of the Methodist New Connexion. Little more was done until 1812 when the Bristol Society for teaching the adult poor to read the holy scriptures was founded by local Quakers and Methodists. The purpose of all the early adult schools was to teach reading only – writing was thought to be unnecessary and possibly subversive, but about 1845 a new generation of schools began to teach writing, arithmetic and other subjects despite strong protests. The Adult School Movement was largely a Quaker organisation and it was under the control of the Friends' First-day School Association which was founded in 1847.[5] With this new objective the number of adult schools now increased rapidly and at the end of the nineteenth century they had their own National Council and a magazine. The Movement still survives and it still has a strong religious bias. Other denominations tried in the second half of the nineteenth century to provide their own alternatives to adult schools and the secular mechanics' institutes. In the North and Midlands these were usually known as Institutes. Sheffield Christian and Educational Institute was founded in association with the Surrey Street United Methodist Church in 1867.[6] At Cambridge the Castle End Mission and Working Men's Institute was established in 1884.† In the South Mutual Improvement Societies were more popular. At Plymouth in 1887 almost every nonconformist chapel seems to have had one.[7]

Adult education of a secular nature and at a higher level began about 1812. Spontaneously in various large towns literary and philosophical societies, institutions or athenaeums were established to provide lectures, a library and a meeting place. In Ply-mouth the Athenaeum or Plymouth Institution was established in 1812 by a group of local reformers 'to promote the cultivation of useful knowledge, by encouraging the habits of research, and affording opportunities to persons of various pursuits to communicate with each other, by the reading of essays on literary and

---

* The Sunday School Movement itself made some provision for 'senior' or 'adult' students (W. H. Watson, *The Sunday School Union*, London, 1869, pp. x, 42).
† The foundation stone was laid by Professor James Stuart – the founder of University Extension – on 6 March 1884.

scientific subjects; and afterwards entering on the discussions to which they might lead'. Seven years later the Athenaeum was able to move to its own building complete with lecture hall, library, museum and scientific apparatus. Early lectures ranged from 'pneumaticks' and electricity to engineering, botany and local antiquities. Despite the total loss of its original building in 1941 the Athenaeum still provides similar facilities in Plymouth.[8] In the same year as the Plymouth Athenaeum was established, the Liverpool Literary and Philosophical Society began with similar aims, and in the next thirty years many other large towns followed their example.[9] There were also more specialised societies of a similar kind – architectural and archaeological, geological or botanical societies. In the West Country there was the Royal Cornwall Polytechnic Society which, like the Royal Society of Arts, offered prizes to 'bring into notice many useful improvements that but for this means would remain unknown'.[10] Most towns of any size had at least one such society in the nineteenth century. They were intended for the middle classes – professional men and manufacturers with sufficient leisure to attend the meetings and sufficient income to be able to pay the annual subscription. The Plymouth Athenaeum admitted 'gentlemen' as members for a guinea a year and 'ladies' as associate members for half a guinea. This was far too much for artisans and mechanics, but was more enlightened than the mechanics' institutions which usually refused admission to women.

Mechanics' institutes were in all other respects intended to be the working-class equivalent of the Societies. The Leicester Mechanics' Institute was formed in December 1833 to provide working men with a library and reading room, evening classes and lectures for 8s a year. Like the Plymouth Athenaeum its supporters were chiefly Whigs and Radicals, and most of its problems arose from disputes between middle-class Whigs and working-class Radicals and Chartists. Eventually the working classes withdrew, leaving control with the Whigs, and the Institute ceased to exist in 1869.[11] In many parts of the country this was the usual fate of mechanics' institutions. Some became little more than places of entertainment, while others like Flora Thompson's in Candleford were reduced to subscription libraries.[12] At Derby in 1883 it was said that the town had a mechanics' institute without a mechanic and a lecture hall without lectures.[13] A fortunate few

3

became centres for adult education and science and art classes subsidised by the central government were held there.[14] In the South and Midlands mechanics' institutes tended to disappear early where there was a middle-class society already and only survived in towns where there was no competition but at the cost of being taken over by the middle classes. The small annual subscription was usually beyond the means of those whom the institutes were chiefly intended to benefit. However they were usually more popular and successful in the North of England. Even such a small village as Kettlewell in Wheardale had its own institute.[15] Separate facilities for women, such as the Bradford Female Educational Institute established in 1857 for factory women, were available. With larger numbers it was possible to form unions. James Hole urged this in his prize essay for the Royal Society of Arts which was published in 1853, and several county unions were subsequently created under the patronage of the Society.[16] The Yorkshire Union was particularly active and it was to play its part in the establishment of university extension courses by Cambridge.[17]

Some mechanics' institutes, notably those at Derby and Nottingham, still survive, while the London Mechanics' Institution is now transformed into Birkbeck College in the University of London. George Birkbeck, after whom the College is named, was professor of natural philosophy at Anderson's Institution (now Strathclyde University) in Glasgow from 1794 to 1804. While there he began a series of lectures for working men in the hope that:

much pleasure would be communicated to the mechanic in the exercise of his art, and that the mental vacancy which follows a cessation from bodily toil, would often be agreeably occupied by a few systematic ideas, upon which, at his leisure, he might meditate.[18]

These lectures were continued after he left Glasgow and led in 1823 to the formation of the Glasgow Mechanics' Institution. Birkbeck moved to London and in the same year of 1823 took part in the foundation of the London Mechanics' Institution with such notable Radicals as Francis Place and Sir Francis Burdett.[19] Birkbeck became the first president of what was to become after many vicissitudes and much opposition the adult education college of the University of London.[20]

The demand for primary adult education was basically religious, but, except for ministerial training, it did not extend far beyond

4

reading and writing. Political motives were much more important in the demand for secondary education. The founders of both the Plymouth Athenaeum and the Leicester Mechanics' Institute, for example, were municipal reformers – men who came to power after 1835[21] – and the same pattern could no doubt be traced elsewhere. Radicals and Chartists, while rejecting the assistance of the Whigs, were equally concerned with adult education. Thomas Cooper, the Chartist, ran classes for his supporters, and the Eight Hour Day Movement demanded '8 hours for work, 8 hours for our own instruction and 8 hours for repose'.[22] In London in 1841 the Chartist William Lovett founded a National Association for Promoting the Political and Social Improvement of the People which proposed to build '*Public Halls* or *Schools for the People* throughout the Kingdom'. These were to train children during the day and be 'used of an evening by adults for *public lectures* on physical, moral and political science; for *readings, discussions, musical entertainments, dancing* and such other healthful and rational recreations as may serve to instruct and cheer the industrious classes after their hours of toil'.[23] But perhaps the most important member of the political Left in adult education was Robert Owen. Owen was born in Wales and became a cotton factory manager in Manchester – where he was a member of the local literary and philosophical society. He bought the mills at New Lanark in Scotland in 1800 to put into practice his belief in Socialism and the Co-operative Movement. In 1816 he opened there an 'Institution for the Formation of Character' – a building which was both day nursery, school, club and adult education centre:

The three lower rooms, which in winter will also be well lighted and properly heated, will be thrown open for the use of the adult part of the population, who are to be provided with every accommodation requisite to enable them to read, write, account, sew, or play, converse or walk about . . . Two evenings a week will be appropriated to dancing and music . . . One of the apartments will also be occasionally appropriated for the purpose of giving useful instruction to the older classes of the inhabitants.[24]

Through Robert Owen the link between the Co-operative Movement and adult education was forged.

The government's part in providing adult education during the nineteenth century, on the other hand, was limited to technical education and motivated by the fear that foreign manufacturers might become more competent technically than British. Michael Faraday summed up this argument in his evidence to the Public Schools Commission in 1862. He felt that it was almost impossible to find 'the intelligent common man' to undertake jobs which required some thought or initiative in England. In France, however, such men could be found easily for quite low wages.[25] For this reason the government opened a Central School of Design in 1837, and in 1841 it began to subsidise similar schools in manufacturing towns.[26] At Leicester in 1849 the mayor welcomed the proposed establishment of a school of design in the town because it would help the local manufacturers to compete with British and foreign rivals,[27] and this attitude to technical education was usual throughout the century. The Great Exhibition of 1851 encouraged it and also provided funds for further expansion. By 1853 the Central School had expanded to become the Science and Art Department based on the South Kensington Museums site and governed by a board.* On the non-technical side the government did nothing until 1851. Some of the schools receiving grants from the Privy Council's Committee on Education had evening classes for elementary instruction attached to them,[28] but no government grant was given for this part of their activities until new regulations were introduced in 1851. Evening classes in reading, writing and arithmetic were then made eligible for the same grants as daytime classes, but it was not until 1891 that further amendments allowed more advanced subjects to be taught.[29] With one exception, to be mentioned in a later chapter, this was the total extent of the central government's involvement in adult education during the nineteenth century.

To return to voluntary activities in the field, both politics and religion combined to found Working Men's Colleges in the mid-nineteenth century. The first was founded by a Congregational minister in Sheffield in 1842. The Rev. R. S. Bayley felt that the local mechanics' institute to which he lectured concentrated too much on practical and scientific subjects, so he founded and ran almost singlehanded the Sheffield People's College with a much wider curriculum. After he moved to London in 1848 the work of the College became more technical and it finally disappeared in

---

* A. S. Bishop *The Rise of a Central Authority for English Education* (1971), pp. 160–6. *The History of the Victoria and Albert Museum* (1952). The inscription 'Science and Art Department' is still to be seen over a side door of the Victoria and Albert Museum.

1874.[30] The London Working Men's College was founded by supporters of F. D. Maurice. Maurice was an Anglican clergyman who belonged to the group of Christian Socialists which included Charles Kingsley and Thomas Hughes. He was dismissed from his post of professor at King's College, London, for heresy in 1853, but his friends and the artisans to whom he had lectured joined together to establish a college for him in Red Lion Square. In 1864 it was joined by a Working Women's College in Queen's Square.* Similar colleges were established in other towns and each provided classes at times convenient for working men and women – in the evenings, in the early mornings and on Sundays. One of these colleges, that at Leicester founded by the Rev. D. J. Vaughan in 1862, eventually became the centre of extra-mural studies of the University of Leicester.[31]

The nineteenth century also saw several attempts to increase the numbers attending university. In 1822 St David's College, Lampeter, was founded to train Anglican clergymen and with the right to grant the degree of bachelor of arts. In 1836 the University of London was formed out of an uneasy union between the non-denominational and Whig University College (founded in 1828) and the Anglican and Tory King's College (founded in 1831). This university was merely a body to confer degrees – a university only in a technical sense, as a later Registrary of Cambridge University was to remark. In 1832 Bishop Van Mildert used part of the revenues of his see to found the University of Durham. Various other colleges which afterwards became universities – Owen's College in Manchester, Mason College in Birmingham, the Hartley Institution at Southampton – were founded about the middle of the century, but there were no further universities capable of conferring their own degrees until 1880 when the federal Victoria University (Manchester, Liverpool and Sheffield) was chartered.[32]

For the higher education of women very little had been done before 1873. Some athenaeums and literary and philosophical societies admitted ladies as associate members and separate women's classes were organised in adult schools and science and art centres in some towns. Mechanics' institutes and working

---

* J. F. Harrison, *A History of the Working Men's College* (London, 1954), pp. 16–21. F. Higham, *Frederick Denison Maurice* (London, 1947), pp. 91–103. In 1866 Maurice became professor of moral philosophy at Cambridge.

men's colleges normally denied women any part in their activities. The Leicester Working Men's College refused to admit them to a singing class where their voices were particularly needed, and later suggested that cookery classes were all that women needed in the way of instruction.[33] Public opinion was against the higher education of women because they were thought to be unfit for it. In 1864 when it was proposed to admit women to London University degrees it was said 'Fears have been expressed that if girls were encouraged to use their brains, the excitement caused thereby would produce insanity'.[34]

The editor of the *Sheffield Independent* believed that the sole purpose of university dons lecturing to young ladies was matrimonial and would consequently lead to an unnecessary increase in that section of the population.*

Those institutions which admitted women tended to segregate them on 'separate but equal' terms. 'It is desirable, of course, as far as circumstances allow, to keep the male and female classes distinct; and this is easily done by means of separate entrances, and by varying the hours and evenings of attendance.'[35]

Elsewhere lectures for women were held in the buildings of the local institute or society but not sponsored by it. At Liverpool in 1838 the pupils of Miss Martineau's school 'paid regular visits to the Mechanics' Institute where a Mr Phillips was giving a course on Pneumatics'.[36] Later in the century a few women's colleges were established. Queen's College, London was founded in 1848 by F. D. Maurice and others – mainly to train governesses. About 1880 the Rev. D. J. Vaughan founded a working women's college at Leicester, and in 1857 Bradford had its Female Educational Institute.[37] But despite these developments it was difficult for any woman to obtain more than an elementary education, and impossible to obtain a degree in England before 1868.

In 1872 Henry Sidgwick wrote from Cambridge that 'female education is in a state of movement just at present here'.[38] This sudden change was the result of a new movement sponsored by women (instead of sympathetic men). They founded Ladies' Educational Associations or Ladies' Lecture Societies in many of the larger towns of the Midlands and the North. Southport had its Educational Association before 1873, Leicester its Reading Society

---

* *Sheffield and Rotherham Independent*, 3 December 1874, p. 2. This was at a time when the celibacy rule for dons was being increasingly criticised.

by 1869, and Cambridge its Discussion Society by an unknown date.[39] They were mainly middle-class organisations because working-class women rarely had the time or the money to join, and the lectures provided were intended for governesses, school-teachers and those who did not need to work.* The existence of these societies, however, did emphasise the need to improve female education. Although there were now more girls' schools than at the beginning of the century, the quality of the teaching in them was not always high and they were mainly private schools available only to the middle and upper classes. There was also an urgent need for higher educational qualifications for women which required the establishment not only of examinations, but also of training courses for the examinations.

Anne Jemima Clough, one of the founders of the Liverpool Ladies' Educational Association, was responsible for the next step in bringing about these improvements. Although brought up in a middle-class merchant's home, her father was later made bankrupt and for some years she supported herself by running a girls' school.† An improvement in her circumstances enabled her to retire early from this and devote the rest of her life to the higher education of women. The appointment of a Royal Commission to enquire into the endowed schools gave her in 1864 an opportunity to press for the opening of all such schools to girls. The advantages of joint lobbying on this and other occasions pointed to the need for some permanent organisation, and led to the establishment of the North of England Council for promoting the Higher Education of Women in November 1867. Miss Clough became the secretary and Josephine Butler (better known for her campaign against the Contagious Diseases Acts) became the first chairman. Delegates from the Leeds, Manchester, Newcastle and Sheffield associations attended the first meeting together with (male) representatives of the Universities of Cambridge, London and Oxford. The Council's purposes were:

(1) to deliberate on questions affecting the improvement and extension of the Education of Women of the Upper and Middle Classes, and
(2) to recommend to the several Associations and Societies herein represented plans for the promotion of these objects.[40]

---

* As one male sympathiser expressed it – 'The time between leaving school and beginning the duties of life was frequently the most neglected, and he feared that was especially so in the case of young ladies.' (*Manchester Guardian*, 11 October 1867).
† Her brother was Arthur Clough the poet.

The first meeting of the Council considered Emily Davies' proposals for the establishment of a women's college (which was eventually to become Girton College, Cambridge), the need for examinations for women teachers, and courses of lectures for ladies (which had already been begun by the four associations acting together).[41] While the Council agreed about the need for a women's college, serious disagreements were to arise between Miss Clough and Miss Davies about its organisation. Miss Davies believed in a Castle Adamant institution isolated from the old universities,* but she did not exclude male lecturers nor did she wish for a separate examination for females. Miss Clough's proposals (which eventually led to Newnham College, Cambridge) were for no segregation from the universities, but a separate examination for women. She was persuaded to adopt separate examinations by her friend Henry Sidgwick who was intent on reforming the degree examinations at Cambridge.

After much debate the Council adopted Miss Clough's proposals and in October 1868 petitioned Cambridge University to establish a higher examination for women. The moment was well-chosen because the Senate was at that time considering extending the permission which it had given temporarily in 1865 for girls to take the Cambridge Local Examinations. The memorial was signed by all the members of the Council – including Thomas Markby who was also secretary of the Cambridge Syndicate for Local Examinations. It was supported by another memorial from resident members of the University.[42] Subsequently the Council circulated its memorial more widely and eventually published it with a list of several hundred supporters from all over Great Britain.† It also issued a questionnaire about the form which the examination should take.

The Senate approved the new examination on 29 October 1868 and the first centres were opened at London and Leeds in July 1869. Out of the thirty-six candidates who presented themselves, twenty-four were successful.[43] Ten years later there were 987 candidates at 21 centres.[44] Despite this rapid increase it was clear that many more would take the examination if teaching were available. At a conference of local secretaries for the Cambridge

---

* W. S. Gilbert's skit in *Princess Ida* (1884) had its basis in serious proposals for a Women's University. See Imperial College Archives, Lyon Playfair MS. 284.

† Printed copy, dated November 1868, in Newnham College Library.

examination held at Leeds in 1871 this point was emphasised and it was suggested that local classes might be organised by the Ladies' Associations or Societies.[45] The North of England Council was in fact already doing this. In the winter of 1868–9 courses of lectures were being delivered in ten English towns and similar courses were available in Scotland and Ireland.[46] Before it dissolved itself in 1875 the Council had done much to improve the education of women in England.

Meanwhile both Emily Davies and Anne Clough had succeeded in establishing women's colleges in association with Cambridge. Miss Davies began at Benslow House, Hitchin in 1869 and not only persuaded Cambridge dons to travel there to lecture, but also persuaded the University to admit her students to the tripos examinations. Unfortunately the successful students were not allowed to take a Cambridge degree, but only received a certificate that they had achieved degree standard.* In 1873 the college was moved to Girton, a mere three miles from Cambridge, and it then became possible for the students to attend University lectures.†
In 1871 Miss Clough and Henry Sidgwick opened a hostel at 74 Regent Street for girls wishing to study for the Cambridge Higher Examination. The following year it moved to Merton Hall at the other end of Cambridge and in 1873 it became a college. Two years later it finally settled at Newnham.[47]

As we have already seen most of the work for adult education was voluntary and almost no assistance was received from the central or local governments. Yet it was clear from the elementary education field that subsidies would be required. As early as 1833 the government began to give grants to the two chief sources of primary education – the Anglican National Society for promoting the Education of the Poor and the nonconformist British and Foreign School Society. Even this proved insufficient and the Education Act of 1870 allowed School Boards to be elected in those areas where voluntary efforts had not provided sufficient school places. The Boards were to be elected by the ratepayers and financed by a local rate. Secular schools were to be erected by them and school attendance was made compulsory for the first

---

* Some of the girls reached a very high standard in these examinations, eventually beating the First Wrangler. *Girton College*, pp. 4–14. Univ. Archives C.U.R. 57.1, 106.

† The girls travelled to Cambridge in 'funereal Girton cabs' pulled by 'the oldest horses in the world' (G. Raverat, *Period Piece*, London, 1960, p. 45).

time. Eventually the Boards were given grants to establish evening classes and it became possible to acquire technical education in the larger towns, but adult education as we know it today was otherwise completely neglected.

It is clear that in 1873 despite great improvements the existing provision for adult education was very inadequate. University extension was a particular need of the day, because the population of the country was expanding much more rapidly than the number of university places. Either new universities must be founded in the larger provincial towns or the old universities must extend their scope to those towns. In 1850 the Oxford University Commission had suggested the latter. In 1855 Lord Arthur Hervey, rector of Ickworth in Suffolk and an active member of the Bury St Edmunds Athenaeum, had published detailed proposals in a pamphlet: 'Why should not University professors, men of the highest standing and attainments, go forth from the Universities and find their pupils in the different towns and cities within a certain distance from them?'*

It remained, however, for James Stuart, a Fellow of Trinity College, Cambridge, to put the plan into practice. From 1867 to 1873 he proved that the proposals were possible by organising courses of lectures for the North of England Council, mechanics' institutes, co-operative societies and athenaeums. He then persuaded the University of Cambridge to take over his work and from 1873 to 1876 he organised the work on behalf of a University Syndicate. His work laid the foundation for university departments of adult education and extra-mural studies throughout Great Britain and a large part of the world. His work also led to the establishment of two universities – Nottingham and Exeter – and encouraged the growth of countless others.

* J. F. A. Hervey, *Lord Arthur Hervey, D.D., Bishop of Bath and Wells* (1896), p. 18. See also H. J. Mackinder and M. E. Sadler, *University Extension, Past, Present and Future* (London, 1891), pp. 3–9. When Bishop of Bath and Wells, Hervey became president of the Wells University Extension Society (*University Extension Journal*, vol. 1, p. 90).

# 2

# James Stuart

James Stuart was a Scot of remarkably varied talents who combined an academic and a political career with his interests in engineering, newspapers and radical reforms. His work for the higher education of women, or for the organisation of training engineers, or for the establishment of adult education would in each case make him an important figure in nineteenth-century history. Perhaps the variety of his interests and achievements have prevented proper recognition of his talents. The *Dictionary of National Biography* inexplicably ignores him. With even less reason D. A. Winstanley, a member of Stuart's own college, omits his name from *Later Victorian Cambridge*, and there is still no biography of James Stuart. His chief memorial in Cambridge is Stuart House, the offices of the University Board of Extra-Mural Studies, but it is unlikely that many inhabitants of the town or the University know after whom it is named. At Norwich the generosity of his sisters-in-law have left several memorials to him and his wife.*

James Stuart was born on 2 January 1843 at Balgonie in Fife-shire, part of the eastern Lowlands of Scotland. His father, Joseph Gordon Stuart, had been trained for the law at the Universities of Aberdeen and Edinburgh, but had chosen a commercial career. He was partner in several flax-spinning mills which had been mortgaged to his father. The family was unusual in several ways and had a strong tradition of radicalism. Napper Tandy, the United Irishman, had formerly been a member of the firm, and James Stuart himself remembered the workmen as 'extreme Radicals, Chartists in those days to a man'.[1] In the mill one of the workmen would read the Chartist newspapers aloud to his workmates. James Stuart's father was apparently not unsym-

[1] References will be found on p. 197.
* Stuart Hall (for adult education lectures), the Stuart Gardens and the Stuart Homes.

pathetic and held liberal views himself. Stuart's mother's family came from Forfar on the opposite side of the Firth of Tay. They were also radicals and friends of the Shelley family. His mother knew Mary Godwin, the daughter of Mary Wollstonecroft, and her husband, William Godwin. *The Vindication of the Rights of Women* and *An Enquiry concerning Political Justice* must have been well known to his parents and influenced James Stuart in turn. In his family political nonconformity was allied with religious nonconformity. J. G. Stuart was a Congregationalist in a country where the only acceptable form of Calvinism was the Presbyterian. James Stuart's religious beliefs were sincerely held, but not in a bigoted fashion. He retained his links with Congregationalism throughout his life, but in order to enter Trinity College and to take his degree at Cambridge he was willing to sign a declaration that he was 'a bona fide member of the Established Church'.[2] His personal beliefs were probably theistic, but this is not entirely clear from his writings.

In his *Reminiscences*, which he wrote at the age of sixty, James Stuart painted an idyllic picture of his childhood in Balgonie. Unlike most English manufacturing towns of the period, at Balgonie masters and men lived in close proximity and met on more equal terms. Education in Scotland was not confined closely to the rich and it was possible for poor boys to get a degree there. Neither was the English distinction between education for gentlemen and education for mechanics so tightly drawn, and it was possible for women to receive a liberal education. To his mother and his grandmother, Stuart owed much of his early education, but one of the family servants gave him his first insight into intellectual pursuits:

I must have been about five years old when he opened my eyes to a whole new set of ideas. He was cutting the lawn in front of the house with a scythe, and, leaning upon it, he said to me, 'How long would you take to count the blades of grass that are growing here?' Of course I could give no answer, but my mind ran to infinities of years. 'I could do it,' he said, 'in twenty minutes;' and he explained to me that he would count the blades in a square inch, and then he would measure the grass, and see how many square inches there were, and so get the number. Looking back through my whole life I do not think that there ever was a single occasion on which I learned so much ...[3]

At the age of eight his maternal grandmother moved to St Andrews and took James Stuart with her. For three years he was able to attend Madras College there. Then, in 1854, he returned to Balgonie and, on the advice of one of his schoolmasters, studied with a tutor. At the same time he took an interest in his father's firm and worked both in the counting house and the mill. The latter, with some engineering drawing taught at Madras College, encouraged his interest in engineering. He also took his first step in adult education at this time by reviving the firm's lending library started by his father. He gave an amusing account of this:

I began by having [the books] re-bound where necessary, and certain missing volumes supplied, and, with the addition of some lighter literature (for the others were all of a seriously instructive character) I opened a library for the village . . .

The thing that I hadn't quite made allowance for was the time they each took to select a book, and, there being only one catalogue, there were some considerable delays. I had some difficulty in keeping them from pushing the stools aside and taking down the books to look at, but I was inexorable on this point, though it seems to me now unreasonably so . . . In the difficulties with which I was beset I bethought me of an expedient. I remembered the reading of the newspapers in the hecklers' shop, and so I suggested to one of my subscribers that he should read the rules aloud, so that they should be well known to all. This gained me just enough time . . .[4]

In 1859 he returned to St Andrews as an undergraduate at the University. In later life he complained of the inefficiency of the professors there:

It will thus be seen that, of all the Chairs in St. Andrews at the time I went to it, there were only four which were filled by professors able to perform their duties. These were Professor Swan (Natural Philosophy), Professor Fischer (Mathematics), Professor Ferrier (Moral Philosophy), and Professor Day (Anatomy). Of these the first two had only been elected to these Chairs the year I entered . . . Of the two remaining, Dr Day, during the course of that session, became paralytic, although he continued his lectures sitting in an armchair in his own house, and Professor Ferrier was suffering so severely from heart complaint that he had frequently to be absent.[5]

Despite these problems he obviously benefited from being at St Andrews, and he paid tribute to the quality of the teaching which he was given. The chief influence on him was the new Principal of the University:

Either the first year I was at St. Andrews or the second year, I am not sure which, Professor Forbes, of Edinburgh, came to be the Principal there. One of the things which gave me quite a new idea was the answer

which he gave to a question of mine as to how he ascertained the proportion that the land in the world bore to the sea. He said he had spread
sheets of transparent paper over the surface of a globe, and marked on
them the contour of the land and the contour of the sea, and then had
weighed the two. It was very simple when you saw it, though John
Richardson's counting the blades of grass was a still better one.[6]

It was Principal Forbes who urged Stuart to enter for the newly-
established Ferguson Scholarship in 1861 and who enabled him
to qualify by persuading the Senatus to dispense with a year's
residence at St Andrews. He obtained the only scholarship for
classics and mathematics, and took an honours degree in classics
the following year. The Principal and Professor Day then persuaded
him to take the examination for a minor scholarship at Trinity
College, Cambridge. He was successful in this too, and in October
1862 he entered the University of Cambridge as an undergraduate.

James Stuart left St Andrews when university reform was only
beginning to affect its life. His was the first honours degree to be
taken there for many years, the first Ferguson scholarship, and the
first scholarship to Cambridge. He reached Cambridge when it
was in the middle of the same reforms and he went to Trinity
College whose fellows were in the forefront of that reformation.
The contrast between the two universities at that time was so
great that it is not surprising that James Stuart paints a gloomy
picture of St Andrews and a glowing one of Cambridge. The
atmosphere of reform and intellectual distinction which he found
at Trinity stimulated him and encouraged those traits in his
character which his early life in Scotland had implanted.

The reformers at Cambridge had concentrated on changing the
administrative system and improving the methods of teaching and
examination. They were assisted in this by the appointment of a
series of Royal Commissions to investigate the working of both
Oxford and Cambridge. In 1856 one body of Commissioners had
replaced the government of the University by the Masters of
Colleges (the *Caput*) with an elected Council of the Senate. At the
same time the Commissioners were given powers to issue new
statutes for the colleges and to transfer funds from the wealthy
colleges to the University, which was seriously short of endowments.[7] A further body of Commissioners established by the
Universities of Oxford and Cambridge Act of 1877 was to carry
these reforms to their logical conclusion while James Stuart was a

fellow of Trinity.[8] The various commissions also made it possible to increase the number of chairs at the University and to improve the methods of appointing to them. Cambridge, like St Andrews, had its share of scandalous appointments. Richard Watson, later Bishop of Llandaff, had been appointed professor of chemistry in 1764 even though he knew nothing of that science. He exchanged it for the chair of divinity in 1771.[9] William Clark, appointed professor of anatomy in 1817, had a paralytic stroke in 1863, but did not resign for another two years.[10] Both these professors were conscientious – Watson became a proficient chemist and a member of the Royal Society afterwards – but the method of election was not satisfactory. Slowly, as money became available to the University in the second half of the century, extra chairs were created and unwieldy subjects divided.

At the same time fresh tripos examinations were created and the old monopoly of classics and mathematics was gradually brought to an end. In 1848 Moral and Natural Science Triposes were established as a qualification for honours. In 1854 an examination in law was set up, while a voluntary theological examination for ordination candidates had begun as long ago as 1842.[11] However none of these examinations replaced the traditional tripos, but merely supplemented it. It was not until after James Stuart's arrival in Cambridge that it became possible to take a degree in subjects other than classics and mathematics. Neither was there a University entrance examination. In 1824 a Previous Examination which all undergraduates had to take in their second year was instituted, but it was not very satisfactory and attempts to improve it had little success. Only one college, Trinity, had an entrance examination for undergraduates.[12]

The University did have a Syndicate which had provided examinations for schoolchildren since 1857. Several organisations, including the Royal Society of Arts and the unions of mechanics' institutes, had campaigned for University-sponsored examinations,[13] but it was Frederick Temple, afterwards Archbishop of Canterbury, who persuaded Oxford and Cambridge to do so. He was at that time a government inspector of schools and had been asked to help in a voluntary system of examinations in Devonshire.[14] Inspired by its success he published a pamphlet on the importance of encouraging middle-class education by a system of examinations and certificates. He followed this up by organising memorials from the larger towns to both Oxford and Cambridge. On 1 June 1857 the Council of the Cambridge Senate reported that it had received memorials asking for local examinations from Cheltenham, Leeds and Liverpool and had also met a delegation from Birmingham. A Syndicate was appointed to consider them and reported in November that there should be one examination for boys under 16 and another for boys under 19.[15] The first examinations were held at Christmas 1858 and proved very successful. Ten years later Emily Davies organised a petition to Cambridge asking that girls should be allowed to take the Local Examinations. This was granted initially for an experimental period of three years, but it was made permanent in 1871.[16] In 1877 it was agreed that the Syndicate's higher certificate should give exemption from the University's Previous Examination.[17] Unfortunately examinations without suitable facilities for education were insufficient to help the middle classes and did nothing at all for the working classes. However the existence of examinations stimulated the demand for teaching and Temple's campaign for the establishment of the Local Examinations Syndicate was later to be followed by James Stuart in his efforts to get a Local Lectures Syndicate. Later still Local Examinations were to provide financial and other support for the work of Local Lectures.

Many of the most active Cambridge reformers were fellows of Trinity at the time that Stuart obtained his minor scholarship there. Henry Sidgwick, afterwards Professor of Moral Philosophy, became a close friend. Sidgwick was at this time greatly concerned about religious tests in the University and in 1869 he resigned his fellowship at Trinity to free himself 'from dogmatic obligations'.[18] But he remained in Cambridge to advise and assist:

I consulted Henry Sidgwick, and gave him the manuscript to read. When he brought it back he had made several corrections, and he gave me a lesson which I have never forgotten in connection with these corrections. He had struck out the words 'There can be no doubt it would,' and inserted 'It would I believe,' and in that, and a few similar instances, he impressed upon me the great advantage of not over-stating a thing.*

Another Trinity reformer was the Bursar, Coutts Trotter, who found Stuart temporary employment writing an official report on endowed schools after he had graduated. Trotter was a lecturer

* *Reminiscences*, p. 170. The manuscript was Stuart's letter to the Senate on University Extension.

on physical sciences and, unusually at that period, had gone to Germany for part of his training.[19] Two other fellows (who were both later to become Bishop of Durham) were J. B. Lightfoot, Hulsean Professor of Divinity, and B. F. Westcott, Regius Professor of Divinity. They did much to further Stuart's career at Cambridge and to promote his appointment as first Professor of Mechanical Engineering.[20]

F. D. Maurice, the first principal of the London Working Men's College and now Knightbridge Professor of Moral Philosophy at Cambridge, was another member of this group. They met in the informal Grote Club (named after Maurice's predecessor) for socratic discussions. James Stuart was secretary of the Club for a time and was able to count on the support of its members within the University.[21] It was the same group of reformers who had done much to change their own college. New statutes which were introduced in 1860 permitted all fellows to take part in the government of the college. Trinity had its own entrance examination for undergraduates and had established minor scholarships – one of the earliest being won by Stuart. It is also to the credit of Trinity that it appointed Aldis Wright as its senior bursar even though as a dissenter he was unable to qualify as a fellow. When James Stuart first went to Trinity the redoubtable Dr Whewell was still the Master. James Stuart only met him once:

Whewell was very much taken with my oration, which was on fairy tales ... [He] shook hands, smiled beneficently, and asked me to dine with him that evening. I went in the seventh heaven, expecting to be introduced as the coming man. I saw in my mind's eye the assembled Dons making an alley for me. Alas! it was very different. I sidled up through a varied multitude. Whewell, who had evidently forgotten all about me, asked how long I had been at college, and didn't wait for the answer.[22]

Until Whewell's death in a fall from his horse, reform at Trinity was slow. Sidgwick noted that it was followed by 'a certain freedom and independence' amongst the fellows.[23] Between 1867 and 1872 a complete overhaul of college administration and teaching was carried out, and it was only the appointment of another Royal Commission which prevented a complete revision of the statutes.[24]

In 1866 James Stuart graduated as third wrangler in the Mathematical Tripos and was immediately elected a Fellow of Trinity. Although a Congregationalist he felt able to make the declaration that he was a 'bona-fide member of the Church of England' for his degree,[25] and a similar declaration for his fellowship. For a time he was absent because of his father's death, but soon after his return to Cambridge he was appointed an assistant tutor at Trinity under the revised tutorial system. He was the first to admit undergraduates of other colleges to his lectures at Trinity:

The experiment was rendered more easy, and more successful, by the fact that the application of mathematics to physical science was beginning to be included in the Mathematical Tripos, and it was precisely to that class of subject that I had given attention. There were few people at that time who had done so, and, in consequence, my lectures were well attended by the students from various colleges who were likely to take a high degree. I lectured on physical astronomy, the mathematical theory of electricity, magnetism and heat, and the application of the potential generally.[26]

In 1875, when the Jacksonian Professor of Natural Experimental Philosophy died, it became possible to divide the duties of the chair and to appoint James Stuart the first Professor of Mechanism and Applied Mechanics. In his application for the post he described his qualifications: 'for a considerable time both before and after I came to the University I was occupied in the practical superintendence and management of machinery, for the last seven years I have given lectures in the University on various branches of applied mathematics . . .'[27]

He held the chair from 1875 until his resignation in 1889. With almost no staff, little accommodation and no financial aid he built up a department of mechanical engineering which attracted pupils from all over Great Britain. One of his achievements was a workshop equipped with machines given by himself and his friends. To pay for the cost of running it, work was done there for other University departments. Elaborate scientific equipment was built for Clerk Maxwell and Coutts Trotter, while Stuart personally designed dust-proof showcases for the professor of zoology.[28]

Unfortunately there was always a group in the Senate which disapproved of the University teaching engineering. When in 1880 Stuart proposed to hand over the workshop to the University they found an opportunity to obstruct. He first offered to sell the machinery which he had bought, but a committee appointed to consider the offer took six years to agree it was desirable. In the meantime he had continued to finance the workshop from his own pocket. After several difficult debates in the Senate his workshop was taken over in January 1887, but almost immediately there was criticism of the way it was run and the way its accounts were kept.[29] Meanwhile Stuart's efforts to get a Tripos Examination in Engineering were being blocked by the Senate. He finally lost patience in December 1889 when a syndicate was appointed to consider the value of the workshop and he resigned his chair.[30] His supporters in the University presented him with a silver salver and an address signed by 190 members.[31]

By 1889 James Stuart was already a Member of Parliament. In November 1882 he had contested a University seat and was defeated by 1302 votes to 3494. Only in Downing College did the number of his supporters almost reach that of his opponent, and he later noted that 'the largest number of clergymen voted against me, I suppose, that has ever voted against any individual person'.[32] It was suggested that he should contest Dundee at the next general election, but before that event the Liberal member for Hackney died and Stuart succeeded him in that safe Radical seat. His election manifesto included votes for women, reform of the Lords, control of public houses, and Free Trade. He had a majority almost as large as his opponent's total vote. As a result of a redistribution of seats he moved to the Hoxton division of Shoreditch for the general election of 1885, and he continued to represent it until defeated in the 1900 election. From 1906 to 1910 he was the member for Sunderland.[33]

Perhaps Stuart's most notable achievement in the Commons was to obtain the repeal of the Contagious Diseases Acts. As early as 1870 Josephine Butler, whom he first met at the North of England Council, had enlisted him in her campaign against the Acts, and he had supported her at various bye-elections on the issue. In 1875 he became joint secretary with her of the British and Continental Federation for the Abolition of State Regulation.[34] Other friends whom he met on the North of England Council led him to support the movement for female suffrage in Parliament, as well as that for female education. He lectured not only for the North of England Council (as we shall see later), but also at both the Cambridge women's colleges. He gave special lectures to Miss Clough's Newnham girls and admitted them to his ordinary University lectures.[35] He travelled to Hitchin to lecture to Miss

Davies' girls too, and later took the much easier journey to Girton for the same purpose.[36] His efforts and those of his colleagues in this field moved James Clerk Maxwell to write two humorous poems on 'Lectures to Women on Physical Science'.[37]

After giving up his chair at Cambridge in 1889 Stuart was reduced to a comparatively small income inherited from his father. To supplement this and to occupy the time left over from his parliamentary duties he turned to journalism. For eight years he edited the *Star* and the *Morning Leader* in the Liberal interest.

At one time I wrote the leading articles in the 'Star', which, though it was an evening paper, required one's attendance at the office at 7 o'clock in the morning. Having rapidly perused the morning papers I then generally wrote the leader at the last possible moment. At first I was terrified with the idea of having only a limited time in which to finish an article, but that soon wore off, and I found it on the whole easier to write against time than not. The paper came out by ten o'clock, and then Mr Parke (the managing editor) and I turned to business questions affecting the undertaking. This generally lasted till lunch time, and having often to attend the House of Commons till midnight I did not get much sleep during that portion of my experience.[38]

While in Parliament he never held government office and his views were probably too extreme for most Liberal politicians of the period. However as a backbencher he was better able to work for the causes which were particularly dear to him. He served on Royal Commissions concerned with the Aged Poor and Local Taxation. The latter led indirectly to more money being made available for adult education, but he does not seem to have spoken much on that or mechanical engineering in Parliament. His knowledge of the latter was probably responsible for his appointment as chairman of one of the Private Bill Committees, because many such Bills give authority for civil engineering works. In 1909 he received the honour of membership of the Privy Council.[39]

Outside Parliament, however, his interest in University Extension remained strong. In 1893, for example, he led a delegation to the Chancellor of the Exchequer to ask for increased grants to the University Colleges, he attended a University Extension Conference at Birmingham, and he opened new buildings at Nottingham University College.[40] In 1889 he was asked to give the inaugural address at the Oxford Summer Meeting of extension students and another to the National Home Reading Union,[41] but after he ceased to be a member of the Cambridge Syndicate he had little to do with its work.

At the end of 1889 James Stuart became engaged to marry Laura Elizabeth Colman, the oldest daughter of J. J. Colman of Norwich. She had been an undergraduate at Newnham between 1880 and 1882 when Stuart still lectured there. Her father was a fellow Liberal M.P.* and a staunch supporter of the University Extension Movement. When the Norwich Extension Society was formed in May 1877 he was on the committee, and in 1897 he became its president – a post later filled in turn by James Stuart, Mrs Stuart and one of her sisters.† Writing to Professor Jackson's wife to announce his engagement, Stuart said:

> I have known them all a long time, and all her people are very pleased and so are all of mine. I had a life of such storm, you see, I will be better of some quiet somewhere in it, if it is to be got . . . The Lady is a good politician; she is a little over thirty; she is engaged in many good works of the same kind as I am. She can make a good speech, and with plenty [of] good sense in it; and I am sure when you see her you will take to her, for she is such a *lady*.†

They were married in the following year at Princes Street Congregational Church in Norwich. When his father-in-law died unexpectedly in September 1898 no one in the family was available to manage the mustard factory at Carrow Abbey except James Stuart. He moved to Norwich for the remainder of his life and was for some years managing director of the works where he established a pensions scheme.[43]

His marriage brought him the quiet he so desired, but he continued to be active in the causes to which he had devoted his life. In 1889 he spoke to the Second Summer Meeting at Oxford about the early days of University Extension. In 1892 Cambridge made tardy amends by asking him to speak at its Third Summer Meeting.[44] There also he spoke of the past work in the field and his hopes for the future. In January 1899 his old University of St Andrews invited him to serve as Lord Rector. In his installation address he once again urged the need for adult education and

---

* He seems to have held the same views as James Stuart on such subjects as the Contagious Diseases Acts, married women's property and the taxation of land values (H. C. Colman, *Jeremiah James Colman*, London, 1905, pp. 359, 370, 400.)

† Trinity Coll. MS. c. 43, 107. Laura Stuart later served on the Norwich City Council and was the first woman J.P. for the city.

training for engineers – the two great works which he had pioneered at Cambridge.[45] Now however he had no need to earn his living as a journalist, and he was able to take long holidays during which he wrote his memoirs. It was at this period of his life that 'he rather resembled a humourous Skye terrier, with his bushy moustache and pincenez balanced crookedly on his nose'.[46] His nieces and nephews remembered him fitting his horses with antlers and arriving at their house in his sleigh dressed as Father Christmas.*

He died at Carrow Abbey on 13 October 1913. A life which had begun in the house beside a Scottish mill ended in another beside a Norfolk factory. Between stretched a life of devotion to good causes, not the least of which was adult education. Trinity College commemorated his work with an inscription in the antechapel composed by Henry Jackson.[47] The University did nothing, presumably because like his successor, Rev. G. F. Browne, it felt that: 'When he did go effectively into politics, he was spoiled by politics ... he fell into the hands of a few extremists of his party, and so long as he voted with them on extreme views, not supported by the real leaders, his influence at Cambridge dwindled'.[48]

Nevertheless when his widow died in 1920 she left £5000 to Cambridge University to endow an extra-mural lectureship in his memory. When new offices were built for the Board of Extra-Mural Studies in 1927 his sisters in-law presented some of the furniture and the building was called Stuart House. In the board-room there his unfinished portrait by Herkomer presides over the work he began a century ago.†

* I owe this episode to his nephew, Mr Humphrey Boardman of Norwich.
† *Reminiscences*, p. 217. The portrait is on loan from Trinity College.

# 3
# The Beginnings

By 1866 when James Stuart took his degree at Cambridge he had already decided on the two reforms which he intended to work for. In the spring of that year while walking with his mother:

I then described to her the great difference that there seemed to me between the education in England and in Scotland. In the former place there were practically no lectures of the kind given by the Professors in St. Andrews, and the opportunities for University education were very much less widespread than in Scotland. I told her that I thought of staying at the University, and of endeavouring to accomplish two things: first, to make the University lectures generally open to all the colleges, and, second, to establish a sort of peripatetic university the professors of which would circulate among the big towns, and thus give a wider opportunity for receiving such teaching.[1]

His origins and Scottish background convinced him of the need for adult education, the spirit of reform and improvement in which he was living at Trinity College encouraged him to propose it, and his friends provided him with the means to carry it out. Lord Arthur Hervey, who had published a similar scheme in a pamphlet ten years before, had been a student at Trinity and still lived nearby as rector of Ickworth in Suffolk. Although James Stuart does not mention meeting him, two copies of his pamphlet were in the College Library and another member of the Hervey family corresponded with Stuart about adult education.[2] James Stuart was brought into contact with the mechanics' institutes by his fellow engineering student, W. R. Moorsom, engineer to the London and North Western Railway Company. That company had its base at Crewe where there was a thriving Institute to which Moorsom lectured. F. W. H. Myers of Trinity introduced Stuart to Miss Clough, the secretary of the North of England Council for the Higher Education of Women. Finally Frederick Temple, then

---

[1] References will be found on p. 198.

headmaster of Rugby, explained to Stuart how he got the Local Examinations Syndicate organised at Cambridge by organising memorials to the Senate.*

His plan was first to organise series of lectures in the larger provincial towns with the aid of his Cambridge friends. Then, when this proved successful, to organise memorials to the University asking them to take over the work. His opportunity came in the summer of 1867. The institutions which were about to establish the North of England Council – the Manchester Board of Schoolmistresses, the Leeds Ladies' Educational Association, the Liverpool Ladies' Educational Association and the Sheffield Ladies' Educational Society had joined together to pay £200 for a course of lectures to be delivered in each of the towns. Although Miss Clough, the organiser of the scheme, must have known of Stuart's interest, she first approached F. W. H. Myers, another fellow of Trinity who had been concerned in the campaign to extend the Cambridge Local Examination to women.† He refused and put forward Stuart's name. Because most of his students were to be governesses and schoolteachers, Stuart was asked to talk on the theory and methods of education.

I replied that I thought it would be much more practicable for me, and more profitable for them, if, instead of an abstract subject of that kind, I simply tried to teach some specific thing in the way, and by the methods, which I might be usefully adopted in other instances, and that, if they liked, I would give a set of weekly lectures in each of those towns on the history of astronomy, eight lectures in each place. They agreed to this.[3]

His first lecture at Manchester was given at Chorlton Town Hall on Thursday, 10 October 1867, with 'a good attendance'. He announced his purpose as not 'so much to give detailed information as to arouse in his hearers a desire to learn something about science'.‡ Beforehand he had been told that the subject of his lectures might be beyond feminine ability to understand or retain, so he adopted a plan of a former professor at St Andrews by providing an outline of his lectures. He included with this a number of

* This was between 1866 and 1869 when Stuart was examining at Rugby (*Reminiscences*, p. 212).
† *Report of the First Meeting of the North of England Council* (1868), p. 13. Myers afterwards became an inspector of schools and president of the Society for Psychical Research.
‡ *Manchester Guardian*, 11 October 1867. It is unfortunate that the Leeds, Liverpool and Sheffield newspapers do not appear to have printed similar accounts of the lectures.

questions to be answered thereby solving another difficulty which had been raised.

Before the course began some people said it would be no use unless there was an examination, and others – these were very early days remember – said it would be unladylike for women to go into an examination. There was like to be a great difficulty, and people came to the first lecture, some intending to denounce the whole thing if the ladies had to stay and go through an examination, especially if it was so young a man who was to ask them questions; and others were ready to belittle the whole affair if I did not ask them questions.[4]

He announced at the first lecture that the questions he had set could be answered in writing and sent to him, but that they were not compulsory. This satisfied both parties.

Only a syllabus or questions, but not both, were given for each lecture on the 'Four pages of large printed octavo' which he prepared beforehand:

## THE HIGHER EDUCATION OF WOMEN

### MR STUART'S LECTURES

Syllabus of Lecture          October 10, 1867

1. The views of Education according to which these Lectures have been arranged.
2. The subjects to be treated of in these Lectures.
3. Late origin of Scientific Enquiry.
4. What is it that renders an enquiry scientific.
5. The process of Scientific Enquiry described.
6. That process illustrated.
7. Four divisions of that process: Observation, Theory, Analysis, Experiment.
8. *Of Observation.* Its function. It supplies facts for theory. Evil of too early introduction of theory.
9. Evil of want of Method.
10. Example. Observations of Meteors.
11. *Of the Formation of Theory.*
12. Of Analysis. Its function. It predicts facts from theory.
13. Extension of the function of Analysis. It also reconciles facts with theory.
14. Impossibility of this reconciliation in some cases. Examples.
15. The kind of Analysis employed in Scientific Enquiry.
16. *Of Experiment.* Its function. It tests theory.
17. Example. The Disentangling of Phenomena.
18. Relation of Experiment to Observation.

19. Restriction of the meaning of the word Experiment.
20. Concluding remarks on these four divisions of the process of Scientific Enquiry. Definition of a Law of Nature.
21. True character of a Law of Nature as apparent from the process of Scientific Enquiry just described.
22. Mistaken views of the Laws of Nature.
23. Correct view of the Laws of Nature.
24. The Laws of Nature inferior to Phenomena.
25. What it is which is superior to Phenomena.
26. Difference in Character between the relations of a Law of Nature to the past and to the future.
27. Concluding remarks on the Discovery of the Laws of Nature.

Questions given at the end of the Lecture    October 17, 1867.

1. How do we know when the sun is nearest to us?
2. What is the evidence from which we conclude that the earth moves?
3. What was the chief objection at first raised to Copernicus' theory? Explain the meaning of the words Down and Up.
4. Give some account of Kepler's life: and point out what you think most worthy of note in it.
5. Kepler's three laws are these:—
   (1) All the planets move in ellipses, having the sun in one of the foci.
   (2) A line joining the planet and the sun sweeps out equal areas in equal times.
   (3) The squares of the periodic times are proportional to the cubes of the mean distances from the sun.
   Explain the meaning of the various technical words used in these laws.
6. What is it that renders an enquiry Scientific? Kepler in some instances guessed at his results:— were the results which he thus arrived at scientific or unscientific?
   . . . . . . . . . . . . . . .[5]

This printed sheet afterwards developed into the printed pamphlet syllabus which Cambridge continued to use until recently. It even includes a sort of booklist at the end although this is clearly for further reading on science in general rather than strictly limited to the subject as later became the practice. This list of eight books includes Tyndal on Sound, the Voyage of H.M.S. Beagle and – the Book of Job. The questions which Stuart set, which also appeared in later syllabuses, were the ancestors of the 'written work' which students are still expected to do.

Although these eight lectures have not survived *in extenso*, James Stuart used the same material for six lectures which he gave at the Crewe Mechanics' Institute in the summer of 1868.

His friend, W. R. Moorsom, had invited him to give a lecture at the recently-opened institute while he was in the North of England. Crewe was a railway town built almost entirely by the London and North Western Railway Company and it provided a very different audience to the ladies of South Lancashire and Yorkshire. However, because the subject of his single lecture was meteors and it immediately followed a display of meteors, he was a great success and more than a thousand people attended. At the end of the lecture James Stuart offered to return and give a course of lectures on astronomy.[6] He gave six weekly beginning on Monday 15 June 1868. The lectures were privately printed at Cambridge University Press[7] so that we have the syllabus, questions and text of the first extra-mural lectures ever given. To the workmen of Crewe he circulated beforehand an open letter setting out his principles in giving the lectures:

Trinity College, Cambridge
May 30, 1868

To the Workmen of Crewe

I wish to take an early opportunity of letting you know what is to be the general character of the lectures on Astronomy which I propose to give at Crewe. I intend to treat of the matter partly from an historical point of view, shewing how the knowledge which we have of the science was gradually brought to its present state; and I shall take care always to explain each step; and, as all Physical sciences are very closely bound up with one another, I shall have occasion often to refer to other sciences, such as Electricity or Magnetism, by way of explanation, and I may even devote one or two lectures to these. For I do not so much wish to present my hearers with any rounded and complete view of Astronomy – which indeed would be the labour of a life-time – but rather to make my lectures as suggestive as possible. For I believe that education generally consists far too much in the learning of simple unsuggestive facts. Now I do not think that there is much use in telling people the results simply to which we have come; the true method is to shew them a road, by pursuing which they cannot help coming to these results for themselves. For instance, it little helps a man towards making a discovery if you tell him what you think is the true method of discovery, but he will be helped wonderfully by learning something of the life of Newton and Kepler, and learning something of what they did, although these things may not bear directly upon what he has in view. There are two things which I wish to avoid; first, giving my hearers a mere dry detail of facts, and second, impressing them with *my* opinion, and not leaving them as far as possible to form their own. I am anxious to do that which is true education, to assist them in forming an opinion for themselves. I will therefore treat with care a few parts of Science rather than wander over it; and I intend therefore in my introductory lecture to give an outline of what I am to do, and I will take care at each step to justify what I say in some of the succeeding lectures; and whilst I unfold some of the grand lessons which I think may be learned, and dwell upon those parts which may best secure the appreciation of my hearers, I will endeavour to point out – or set them on the way to find – other and perhaps better examples . . .[8]

This open letter sets out very fully and accurately the principles which were later to govern extension lecturing. It is interesting to note that for a few years at least the early Cambridge lecturers tried to get similar open letters to their potential audience published in local newspapers.

It was Josephine Butler who drew James Stuart's attention to the possibility of co-operation with the co-operative societies. She and other feminists had been promoting a Married Women's Property Bill, one of its purposes being to protect the right of wives to retain the 'dividends' declared by the societies. During the research for this she discovered that many societies put aside part of their profits for educational work, but found great difficulty in obtaining good lecturers. She put them into touch with Stuart who agreed to give a course of lectures.[9] In 1869 he lectured for the North of England Council again, and also for the Rochdale Equitable Pioneers' Society – the first co-operative society to be founded on the modern plan. To the ladies of Liverpool he gave nineteen lectures on natural philosophy. A summary of each of the first eight lectures and the questions which he set on them were printed 'for private circulation',[10] but the other eleven were not preserved in this way. At Rochdale he gave a shorter course on the same subject, and it was there he established another principle of extension lecturing:

It was at Rochdale that the plan of having a class in connection with University Extension Lectures originated. One day I was in some hurry to get away as soon as the lecture was over, and I asked the hall-keeper to allow my diagrams to remain hanging until my return next week. When I came back he said to me, 'It was one of the best things you ever did leaving up those diagrams. We had a meeting of our members last week, and a number of them who are attending your lectures were discussing these diagrams, and they have a number of questions they want to ask you, and they are coming tonight a little before the lecture begins.'[11]

For many years afterwards, Cambridge courses had a lecture and a class and not all of those attending the lecture also went to the class. The distinction has been abandoned for many years now, but traces of its influence still linger on.

James Stuart not only established the pattern of extension lecturing in two years. He also proved that there was a demand for courses of lectures to a university level, and that young Cambridge fellows could be induced to give them. Only with the co-operative movement does he seem to have failed.* In 1870 he began the second part of his campaign for university extension. This was intended to persuade Cambridge University to take over the work which he had begun, put it on a permanent basis, and extend it throughout the country. Just as the Local Examinations Syndicate had been established in 1858 by pressure from the larger provincial towns carefully organised by Temple, so Stuart began to get memorials sent to the University in favour of his plan. From 1871 he left the courses to his Trinity friends and toured Northern England and the Midlands lecturing to any interested body who could assist his plan. He also printed this lecture as a pamphlet – *University Extension . . . Being the Substance of an Address delivered first at the Request of the Leeds Ladies' Association.*† In it he differentiated between primary education, technical education, and higher education. He defined the last as teaching people to think, and considered it the main task of the university extension lecturer.

It has frequently fallen to my lot to examine, for instance, the Euclid done at night schools. I have found lads who wrote out the propositions of Euclid excellently, but who were, nevertheless, perfectly unable to apply these propositions to any practical problem, such as measuring the breadth of a room or the height of a tree. They had acquired, as it were, a certain amount of information about Euclid, but they had not been taught to think about it. It is very much easier to make a person learn a thing than to teach him to think about it.[12]

Only the universities could provide this kind of education and they had already started to extend their teaching by admitting non-collegiate students.‡ James Stuart hoped that this would soon be followed by the admission of women, but considered that the universities would still be unable 'to reach the masses of our

* Most of the societies seem to have paid lip service only to adult education both then and later, but there may have been difficulties which Stuart did not know of.

† Leeds, 1871 (BEMS 1, 11). It was delivered at a conference organised by the Leeds Association on 26 April (Leeds City Archives, Yorks. Ladies Council, vol. 56A.)

‡ Fitzwilliam House was opened at Cambridge for non-collegiate students in 1869. It became Fitzwilliam College in 1966.

countrymen and countrywomen ... who are eager for education and cannot get it'. He therefore proposed a peripatetic university which should be endowed partly from funds raised locally and partly from university and college endowments, just as Lord Arthur Hervey had done in his pamphlet published nearly twenty years earlier. Stuart's lectures not only encouraged the preparation of memorials to Cambridge, but it also stimulated the demand for university extension lectures. His lecture at Rochdale in June 1871 for example led to a letter from a working man asking for advice on studying 'when they have not the aid of a teacher'.[13] By the beginning of 1872 he had aroused interest in Birmingham and Yorkshire. The Yorkshire Union of Mechanics' Institutes invited him to speak at their annual meeting in May when his lecture was interrupted by 'bursts of emphatic applause'.[14] He reported progress to his mother on 13 March 1872:

At Birmingham they have formed a committee and are preparing a memorial to the University from that place and from the midland districts. I expect a memorial *may* come from the Yorkshire Board of Education. It consists mostly of *Gentlemen* but there is a more valuable source from which I also expect a memorial namely the Central Committee representing all the mechanics institutes of Yorkshire. There is a very important man connected with these Institutions a Mr Curzon who is the general secretary and I am very desirous of seeing him about the question ...[15]

Stuart also addressed the newly-formed Chesterfield and Derbyshire Institute of Engineers in May and in June a public meeting at Leeds on the occasion of the North of England Council's meeting in the town. In July he was invited to speak to the Co-operative Conference on Education at Bury in Lancashire, and the *Co-operative News* took up the cause enthusiastically, printing the greater part of his speech in their columns.[16]

By these and other means James Stuart got in touch with men and women in the provinces who were committed to the cause of adult education. The lecture at Leeds put him in touch with the Rev. J. B. Paton, a Congregational minister at Nottingham. Paton was a supporter of the local mechanics' institute and had also founded an 'Institute' for the training of Congregational ministers.* In later life he wrote:

When in London I met Sir Edward Baines, who told me, to my surprise and delight, that Mr James Stuart, Fellow of Trinity College, had, the

* The Institute later became Paton Congregational College, Nottingham. It was amalgamated with Manchester Congregational College in 1969.

week before, delivered a lecture in Leeds in which he had specially urged the duty of the great National Universities to undertake some extension work amongst the people of England for Higher Education. Within an hour of learning this glad news I wrote to Mr Stuart, and then the association was formed between Nottingham and the University of Cambridge.[17]

Although Paton says that this happened in June 1873, his memory was at fault and it took place twelve months earlier. As early as January 1873 Paton and a local Anglican minister, Canon Morse, had persuaded the local mechanics' institution to petition Cambridge University:

To the Vice-Chancellor, Council, and Senate of the University of Cambridge. From members of the Mechanics' Institution, Nottingham, in Annual Meeting assembled.

Gentlemen, – Believing that many circumstances conspire to make the higher education of the people of England, and especially the industrial classes, a matter of momentous importance, and that the ancient Universities are able to take the guidance of this higher education, so as to become, in the truest and most emphatic sense, National Universities, we beg to submit to you the following facts ... Information has been received that proposals similar to those above mentioned have been laid before your University from Birmingham, Crewe, Leeds and Rochdale, and the inquiry of which the Council of the University has promised to make into the possible co-operation of the University into the highest education of the people in our large towns. This meeting begs to assure you of its grateful sense of the promptitude and sympathy with which the University has promised this inquiry, and desires to suggest that the following proposals should be carried into effect: –

1. To appoint lecturers of approved eminence and skill, who may conduct evening classes for working men in our large towns, and also at other times give lectures to the more educated in the same localities, so as to spread the advantages of University education throughout the country, and to all classes.

2. To make arrangements with the various towns soliciting its aid, so as to divide them into such circuits as will engage the full time of a lecturer, and afford him adequate remuneration; and

3. to give distinction to those men who shall thus devote themselves to the higher education of the people by admitting them to share the titles and privileges of fellows in the colleges ...*

At the neighbouring and rival town of Leicester, James Stuart was probably already acquainted with the Rev. D. J. Vaughan, vicar of St Martin's church. Vaughan had himself been a fellow

* *35th Annual Report* of the Nottingham Mechanics' Institution (1873), pp. 19, 20.

of Trinity College and was still a constant visitor to Cambridge.* Vaughan had come under the influence of F. D. Maurice's Christian Socialism and was a close friend of J. L. Davies, a founder of the London Working Men's College.[18] In 1862 Vaughan founded the Leicester Working Men's College in imitation of it. Because of his interest in adult education, Vaughan knew the local Quakers who had organised a strong Adult School Movement in the town. The chief family was the Ellises who lived at Belgrave Hall and were also founders of the Midland Railway Company. That company had its headquarters at Derby and so through Canon Vaughan, Stuart was introduced to potential supporters in both towns.[19]

Public meetings were held in all three Midland towns in 1873. At Nottingham the mayor summoned a meeting on the afternoon of Monday 7 April. After explanations of what was proposed, resolutions in favour of joint lecture arrangements with Leicester and Derby were passed and a committee of ladies and gentlemen was elected to carry on the work. Negotiations between Leicester and Nottingham representatives had obviously already taken place and the Rev. Francis Morse of Nottingham was corresponding with the Rev. Mr Vaughan in Leicester.[20] James Stuart was not present at this meeting, but he attended the second of two held at Leicester in April. This was in the Mayor's Parlour of the Guildhall on the evening of 25 April when Stuart, Vaughan and others recommended the scheme. The local Liberal newspaper concluded its account of the meeting with the wish that 'Our readers will no doubt receive more information concerning this subject'.[21]

Derby, the third of the North Midland industrial towns, did not hold its public meeting until May. The mayor issued a handbill calling a meeting for the evening of Friday 2 May to be addressed by Stuart on a scheme for the instruction of

1. Young Men of the Middle and Upper Classes.
2. Ladies –
   (i) Those who have left School.
   (ii) Governesses and Schoolmistresses.
3. Working Men.[22]

The Rev. Walter Clark, headmaster of the local grammar school, wrote to the local newspapers on the importance of the subject,

* Leicester University, Vaughan College attendance register 1869–95 records some of Vaughan's visits to Cambridge. His church is now Leicester Cathedral.

but only one printed his letter.[23] Stuart, Morse and Vaughan all sent apologies, but Clark was well informed about the scheme and was even able to announce the proposed subjects for courses. A committee was elected to make all arrangements. This included a local manufacturer, T. W. Evans, who had graduated at Cambridge and was to prove an enthusiastic local supporter.[24]

Within the University James Stuart began to seek public support in his campaign in 1871. On 23 November of that year he issued as a flysheet

A Letter on University Extension Addressed to the Resident Members of the University of Cambridge.

Gentlemen,

In the few following remarks I beg to call the attention of the University to one phase of University extension, and to urge the desirability of members of the University considering the subject both in their corporate and in their individual capacity.

We have of late, by permitting the residence of non-collegiate Students, taken a great step towards rendering our Universities more accessible to all classes. By that step the Universities may be truly said to have opened their education to all men. I do earnestly hope, and I do not doubt, that they will also ere long open their education, and their endowments to all women, but 'all men' and 'all women' in these sentences only include those classes who can procure some years of continuous leisure, which is far harder to be got than the requisite money; and there must always be a vast multitude of persons who cannot command that continuous leisure. Amongst those classes whose circumstances inevitably debar them from residing at an University, there exists a wide-spread desire for higher education of a systematic kind. I am not referring to any spasmodic desire, but to one accompanied by a thorough understanding that continuous personal effort on their part is requisite in order to acquire education, and accompanied also by a willingness to undertake real labour for that end. I would not venture to make this statement, nor to bring this question before the University, if I had not by a considerable experience, now extending over five years, become convinced as well of the thoroughness as of the existence of this desire. Now, the desire indicated is an excellent one, and should not be allowed to waste itself away for want of assistance. When these people cry for bread, a stone should not be given to them, as is too frequently the case with those popular lectures which are got up by Mechanics' Institutes and the like.

These and Night-schools have been established, owing to the necessity of oral or personal teaching, which can never be superseded by anything in the whole province of education. The point at which such Lectures and Night-schools are at fault is different in each. True education must be continuous, and must be given by a man of thorough attainments. It has as yet been generally impossible to unite both of these qualifications because of want of funds. Thus we find on the one hand (with certain brilliant exceptions) that those teachers who are attempting to give education of a higher kind continuously to the working classes are men of insufficient attainments, for, whatever be the case in individual instances, the class of teacher will be determined by the amount of salary offered to him; and on the other hand, that when men of high attainments are obtained, their teaching is in general too discontinuous to deserve the name. Isolated pieces of instruction can never be truly educational ...

There are many methods according to which such a plan could be carried out. It should be the duty of a man so appointed to devote his time to teaching by Lectures, and also to the very important work of Night classes. There are, perhaps, few towns just now which could support a permanently resident staff of Professors, and much education may be well disseminated at present by less ambitious effort. Many places which could not with advantage have several teachers at once, could well take them successively, and an interchange of teachers occasionally between various towns willing to co-operate in such a scheme would secure the teaching in each town of a due variety of subjects. I have more than once been led by experience to see that the residence in this or that town, even for one winter, of a man of recognised attainments, whose duty it would be to devote himself to teaching the subject to whose study he had applied himself, would be of an immense benefit to that town, and would readily meet with appreciation in a place where perhaps more extended effort would always remain impossible unless some such influence as this has prepared the people for it.

In some large towns, however, we might well establish a small permanent University, the Professorships of which, perhaps three or four in number, might be endowed by one or more Colleges, and held by such Fellows of those Colleges as might be willing to retain their Fellowships in virtue of the performance of such work. The appointments and the regulation of such branch University should be entirely in the hands of those who founded it. In such circumstances we might well call upon the town to contribute the buildings, etc., and to bear all incidental expenses, nay, even to assist in providing the Endowments.

The two methods I have here indicated, that of a peripatetic and that of a fixed college, might to some extent be united. And, in carrying out a system of an interchange of teachers between various places, an occasional interchange between the University and the towns might be productive of beneficial effect to both.

The education which such a teacher should give should be open to all persons, but should be directed more especially to those classes who most desire education, and whose circumstances are such as most to prevent them getting it. The classes who have most expressed that desire, and who have made most exertions to obtain education, and to take advantage of what has come to their hand have undoubtedly been women and working men. But, while his efforts would at first be most directed to those classes which responded most, the very existence of such a system in their midst would arouse and stimulate the desire for education in a variety of other quarters. But as he would find these classes, so to speak, ready to his hands, constituting a nucleus in many instances already formed from which to commence operations, he might at least begin from these and give a course of lectures once, or perhaps even twice, a week, in the evening to the working people, and take a night class twice a week in the same subject as that on which he was lecturing. He might also give a course of lectures, say twice a week, for women, with perhaps two hours a week for class teaching in connection with these; all such lectures should be accompanied by syllabus and questions to be answered in writing, and by conversation before and after the lecture, with a view to suggesting or solving difficulties. In addition, however, to women and working men, a teacher would undoubtedly find others, and these, owing to his presence, a rapidly increasing number, who desire instruction; for these two lectures a week with two hours' Class teaching would probably be sufficient. It would be a great saving of trouble if the lectures to men and to women were not separate . . .

I have taken the present opportunity of bringing this question before the notice of the University, inasmuch as one or two memorials from bodies practically interested in education in large towns have been or are about to be forwarded to the Vice-Chancellor and Senate. In any scheme we must be prepared to meet with disappointment and in some quarters with want of response. But I would not have written this letter if I had not believed that there are certainly places where such a system as I have indicated would be at once responded to, and from which it would rapidly spread. But, while I have based my arguments on the existence of a demand, I at the same time believe that it is not only our duty to foster and encourage it where it exists, but, by the attitude which we assume, to endeavour to call it up where it does not exist, or has not had energy enough to express itself. Enquiry and discussion however can alone bring to the surface a clear knowledge of the quarters in which effort should be first begun, and the character which that effort should have. The memorials to which I have referred ought to indicate some of the directions in which our efforts would be useful and in which our enquiries might be directed; and I would strongly urge that the question of how we might afford assistance to the higher education in great towns of those classes who are inevitably debarred from residence at an University, should be referred either in a general or in a particular form to the Local Syndicate, or some Syndicate appointed for the purpose, in order that they may report upon it to the Senate, and suggest what steps, if any, the University or Colleges might in their opinion most advantageously take.

JAMES STUART[25]

The memorials to which he referred were from the Rochdale Equitable Pioneers' Society, the Crewe Mechanics' Institute, the North of England Council for the Higher Education of Women, and a public meeting at Leeds. The memorial of the Rochdale

Co-operative Society explained that it devoted 2½ per cent of its profits to educational purposes. It had provided a library of 10 000 books in the town and had organised lectures which 'have, we believe, been of use to those who attended them in encouraging a taste for literary or scientific pursuits, but have in general had the fault of being too desultory'.[26] If the University could provide good lecturers for courses on different subjects, the Society would provide accommodation, equipment and a subsidy. The Crewe Mechanics' Institute described their work for primary and technical education:

In addition we have occasional Lectures delivered by such Lecturers as voluntarily come within our reach and on subjects which are only occasionally connected with those taught in our Classes. These Lecturers do not in our experience produce so permanent an impression as we would desire; they are frequently popular in tone and aim rather at providing rational amusement than at arousing the mind to active exertion. From these causes and from the want of continuity in the Lectures delivered in any one Session, and the absence of any means of following up and making permanent whatever slight impression may be produced by the Lecture, it results that we have not made as much progress in the mental and moral improvement of our population as we might have done under better auspices.[27]

The memorial goes on to record the success of Stuart's course at Crewe and to ask the University to organise further courses.

The North of England Council in its memorial drew attention to the courses of lectures for women which it had organised over the past five years. These courses had been chiefly intended for schoolmistresses and 'ladies who have left school and who can advantageously pursue their studies in this manner without the necessity of leaving their homes'.[28] The Council pointed out the financial and other difficulties involved in organising these courses and their failure to include many classes of men and women who might benefit from them. The memorial from Leeds was signed by the mayor, the chairman of the school board, a local M.P. and influential townsfolk.

The Memorial of the undersigned Inhabitants of the Town and Neighbourhood of Leeds in the County of York
Sheweth,
    That for some years past there has been in the Town of Leeds as well as in other great centres of population in the Kingdom a great and growing desire for some means of higher education accessible to those who are unable to reside at a University.

38

That in the Town of Leeds, which contains a population of more than a Quarter of a Million, and in the adjacent Towns and Villages forming the manufacturing District of which Leeds is the Capital, there are large numbers of young men engaged in almost every kind of professional and business occupation, while so few of these have graduated at a University, that they only prove the fact that the Universities have little or no direct influence on the professional and mercantile classes of the District.

That the number of young men who go from Leeds to a University is very small, and of these a large proportion do not return but take Orders in the Church or are called to the Bar, or adopt some other profession which calls them to live elsewhere, so that of the small number of men who go from the district to a University a small proportion only return to reside there.

That with very few exceptions the men who form the professional and mercantile population of Leeds leave school at ages varying between 16 and 19 years, and at once begin to learn a profession or business; few of them have the means requisite to enable them to proceed to a University, and others are deterred by the consideration that to proceed to a University would be to postpone for 2 or 3 years the time when they would begin to acquire the special knowledge required to fit them for their future career, and the time when they would be able to earn their livelihood.

That the time devoted to the acquisition of technical knowledge during the first years of professional or business life in most cases allows leisure for the continuance of general culture, and many youths would gladly avail themselves of facilities for keeping up and extending the knowledge acquired at School.

That at present no facilities of this kind exists beyond series of isolated Lectures and unguided and desultory reading which is within the power of many, but which few have the perseverance to continue or the ability to make profitable.

That the Universities have conferred a great benefit upon the Inhabitants of the Country generally by the introduction of Local Examinations which have brought the influence of the Universities to bear directly on many who formerly derived no benefit from them, and the more recently established systematic courses of Lectures to Women by University men have further extended the influence of the Universities and been productive of the greatest benefit to those who have attended them.

That the formation of systematic Courses of Lectures and Classes on a permanent and broad basis in the large centres of population would confer a great benefit on the Inhabitants and supply a want which is generally felt.[29]

The four memorials were printed by order of the Vice-Chancellor on 30 January 1872 and circulated to members of the Senate. James Stuart hoped that the result would be the appointment of a Syndicate 'before the middle of April so that the University may be able to act about them before the end of that term and so not to have to defer the matter over the Long Vacation'. He had the support of many influential people in the Senate including the two professors of divinity (Westcott and Lightfoot), but thought that he needed more memorials from bodies which the University would listen to.[30] In April he hoped (unsuccessfully) for memorials from the unions of mechanics' institutes in Lancashire and Devonshire, but by May four more had been received by the Vice-Chancellor. These were from a public meeting at Birmingham, the Leeds Mechanics' Institute, the Ladies' Council of the Yorkshire Board of Education, and the Yorkshire Board's governing body.[31] Only the Birmingham memorial was printed in full although a subsequent report to the Senate summarised the contents of the others. The Birmingham memorial deplored the lack of opportunities for higher education and named three conditions for its introduction:

Firstly, a standard of excellence in the principal departments of Literature, Science and Art fixed by some universally recognised authority, and attainable by students of this class, which would secure for their studies the definiteness and thoroughness that are so much needed.

Secondly, an opportunity, offered to all who might be inclined to take advantage of it, of bringing their acquirements to the test of an Examination.

Thirdly, the command of teaching power of a high order for the benefit of those who might wish to place themselves under instruction.[32]

Instead of appointing the Syndicate which James Stuart desired, the Council of the Senate decided to send a letter to each of the memorialists accepting the need for action and asking a series of questions:

1. What demand is there for instruction in your locality among
   (1) Young men of the middle classes who have left school,
   (2) Women of the middle classes who have left school,
   (3) Young persons and adults of the working classes, (a) Men, (b) Women?
2. What provision is there in your locality for affording instruction to such persons?
3. What associations are there in your locality connected with the education of these persons?
   Would they be willing to co-operate with the University in a scheme of instruction by lectures and classes?
4. What means are there for obtaining the co-operation of working men in organising lectures and classes?

5. Can you mention any other places which would be likely to co-operate with you?
6. What times of the year would be most convenient for lectures and classes? . . .
7. What days and hours would be most convenient for the above purposes?
8. How many hours a week could the youths and the adults of the working classes in your locality devote to attending lectures and classes?
9. What subjects do you think the persons in your locality would chiefly desire to study?
10. How far do the classes and examinations in connection with the Science and Art Department of South Kensington supply the educational wants of your locality?
11. Do you think that University examinations alone, without University teaching, would meet the wants of your locality?

Replies were received from most of the memorialists and also from two more Lancashire co-operative societies, the Ladies' Educational Association of Leamington and other interested bodies. James Stuart not only sent out further copies of the questionnaire, but also helped to compose the replies. The draft of the answers from the Crewe Mechanics' Institute is to be found amongst his papers,[33] and the returns from Derby, Leicester and Nottingham were very similar. The replies from Nottingham were printed in full by the University:

*Answer* 1. In the town of Nottingham there is a population of about 87000, and in the immediate suburbs an additional population of about 55000, making on the whole about 142 000.

The wages of the operatives range as follows: –

Lace Makers from £1 10s. to £2 10s. per week.

Hosiery workmen and other skilled artizans from £1 5s. to £1 15s. per week.

Women from 8s. to 16s.

The incomes of the manufacturers, merchants and tradesmen bear a high average.

It cannot be doubted that there is a large sphere of usefulness open to those who will undertake such teaching amongst the following:

1. Young men, sons of manufacturers, professional men, and tradesmen, who have been taken early from school and put into business, but have no occupation for their leisure hours, and are consequently liable to the temptations of billiard rooms, theatres, etc., and even if desirous of self-improvement have no encouragement or helpful guidance, such as they would find in classes conducted by University teachers.

2. Young ladies who have left boarding and other schools, and have their whole time unoccupied . . .

3. The young men and women of the working classes, many of whom now spend their evenings in walking about the streets or at places of public amusement, theatres, music-halls, dancing-rooms.

4. The elder men of the working classes; many of them now spend a great part of their leisure time in public-houses, where there is much discussion of national and local political questions, and of matters connected with trade and manufactures.

The artizans of this locality are especially distinguished for technical skill, but do not understand the scientific principles applicable to arts and manufactures. Much help would be given to them if these were efficiently taught.

*Answer 2.* The only provision for the instruction of such persons is afforded : –

1. *In Art,* by the Government School of Art . . .

2. *In Science,* by the Government Science Classes . . .

3. *In General Reading,* by the Free Library containing 19 000 volumes, the Library of the Mechanics' Institution containing 7800 volumes, the Library of the Peoples' Hall containing about 3200 volumes, and two 'Operatives' Libraries' kept at Public Houses, containing about 3000 volumes.

4. *In Languages,* by the classes of the Mechanics' Institution . . .

5. *In Literature and Science,* by the Lectures (to members only) of the Literary and Philosophical Society, by popular lectures at the Mechanics' Institution, and by the discussion classes connected with various places of Worship, but none of these provide continuous teaching.

*Answer 3.* There are no associations having for their object the education of such persons, except the Committees of the Government Art and Science Classes: no doubt these Committees would be willing to co-operate in a scheme of systematic instruction as far as practicable.

*Answer 4.* Representatives of some of the most prominent of the local trades have joined the Committee, and will be best able to obtain co-operation of those whom they represent.

The interest of one of the trades, namely the 'Machine Makers', in improved teaching has been shown by their giving two prizes to the Government Class for Machine Construction and Drawing.

*Answer 5.* Leicester and Derby, also the populous Iron and Colliery District of the Erewash Valley 10 miles from Nottingham, are likely to co-operate with the Nottingham Committee.

*Answer 6.* It would be easy to form classes from the end of December to the end of January (avoiding Christmas week); also for three weeks at Easter, if Easter did not fall very late.

Between the middle of June and the 1st of October would be the worst time in the whole year.

*Answer 7.* The most convenient hours would be, for ladies' classes, 10 A.M. to 1 P.M. or 3 to 5 P.M., on any day of the week.

For young men of the upper classes, 7.30 to 9.30 on any evening except Wednesday and Saturday; but probably they could generally meet between 3 and 5 P.M. on two afternoons (other than Saturday) in each week.

For the working-classes 7.30 to 9.30 on any evening except Saturday.

*Answer* 8. They could without difficulty attend a lecture one evening and a class another evening in each week, say three hours per week . . .

*Answer* 11. Certainly not.[34]

## A Syndicate was finally appointed on 27 February 1873:

to consider the subjects of the Memorials from Birmingham, Leeds, Nottingham, Crewe and other places, requesting the cooperation of the University in establishing a system of higher education in various parts of the country, and to report to the Senate before the end of the Easter Term 1873.*

The members of the Syndicate included the Master of Trinity, Professor Westcott and Professor Lightfoot, Henry Sidgwick and James Stuart himself. They deliberated and they met 'deputations from the great centres of population'. In a letter to his mother James Stuart was overjoyed – 'There never was such a deputation at the University before – fancy the Masters of Colleges and representatives of trades unions laying their heads together to negotiate a means of raising the state of the people's Education!'[35] The Syndicate reported in favour of a two year experiment on 24 May and its report was accepted by the Senate on 5 June. James Stuart felt that it was the culmination of five years of effort even though the University would not provide a subsidy: 'it would be impossible to bring the matter to a successful issue, if the Scheme were to involve any proposal for a charge upon University or College funds. The system of Local Examinations, established a few years ago, which has been regarded as a great boon to the Country, is entirely self-supporting.'[36]

This was unfortunate because it was soon to become obvious that local lectures could not be self-supporting as local examinations were, but without this concession it is unlikely that the proposals would have passed the Senate. The report and its appendices form an important and historic document. It describes in some detail the course of twelve lectures and classes which was to be a standard in university extension work for many years. It anticipates the establishment of permanent university extension

* BEMS 4/1, 1. This Syndicate also reported on the proposal to establish a County College at Cambridge. This was established as Cavendish College in 1876, but it was closed in 1892 (*Victoria County Hist. of Cambs.*, vol. III, 1959, p. 266).

colleges for the large centres of population and intermittent courses for the smaller towns. It foresaw the examination which would provide a qualification for those who had attended a series of courses. All these suggestions were brought together by James Stuart and he had the satisfaction of seeing them in practice during the next few years. Until Albert Mansbridge invented the three year tutorial class forty years later almost all university extension work followed Stuart's plan. Out of the proposal for colleges eventually came the University Colleges of the early twentieth century – Nottingham and Exeter being the children of the Cambridge Syndicate. In 1879 the first Vice-Chancellor's Certificate was issued to a woman student at Sheffield for her success in examinations over three years. Stuart also had the satisfaction of seeing his experiment imitated by other universities. A London Society for the Extension of University Teaching was founded in 1876 and later was incorporated in the University of London. In 1878 the University of Oxford made a timid start with a Delegacy for the same purpose. Most of the other universities founded during his lifetime were not slow to follow the example of the Cambridge Syndicate.

# 4

# The Local Lectures Syndicate
## 1873-6

So certain were James Stuart and his supporters that they would be successful that while the Senate was debating the Syndicate's report, they were making arrangements for courses of lectures to begin in the autumn. A delegation from Derby, Leicester and Nottingham met the Syndicate on the day of the debate to discuss the arrangements and subjects.[1] On 13 June James Stuart produced their plan to the Syndicate. In the first term of three months there was to be a course on Political Economy 'intended specially, though not exclusively, for the working classes'.*

| | |
|---|---|
| Derby | Lectures at 7.30 p.m. on Tuesdays |
| | Classes at 7.30 p.m. on Fridays |
| Leicester | Lectures at 8 p.m. on Wednesdays |
| | Classes at 8 p.m. on Saturdays |
| Nottingham | Lectures at 7.30 p.m. on Mondays |
| | Classes at 7.30 p.m. on Thursdays |

The second course in the Michaelmas term was on English Literature and was to be 'at a time which it is expected may generally be found most convenient for ladies'.†

| | |
|---|---|
| Derby | Lectures at 11 a.m. on Wednesdays |
| | Classes at 11 a.m. on Fridays |
| Nottingham | Lectures at 11 a.m. on Tuesdays |
| | Classes at 11 a.m. on Thursdays |

The third course on 'Force and Motion' was intended for 'young men engaged in Business'.‡

| | |
|---|---|
| Derby | Lectures at 7.30 p.m. on Wednesdays |
| | Classes at 7.30 p.m. on Saturdays |

[1] For References see p. 198.
* BEMS 4/1, 55. The lecturer chosen later was the Rev. V. H. Stanton of Trinity.
† BEMS 4/1, 55. The lecturer chosen later was the Rev. E. B. Birks of Trinity.
‡ BEMS 4/1, 55. The lecturer chosen later was T. O. Harding of Trinity.

45

Leicester     Lectures at 8 p.m. on Thursdays
                Classes at 8 p.m. on Mondays
Nottingham  Lectures at 7.30 p.m. on Tuesdays
                Classes at 7.30 p.m. on Fridays

For the Lent term Astronomy took the place of Political Economy for working men, Physical Geography succeeded English Literature for ladies, and English Constitutional History followed Force and Motion for the young men. All other arrangements remained the same, except that Leicester had lectures in Physical Geography at 11 a.m. on Tuesdays and classes at 11 a.m. on Fridays.[2]

Written Examinations will be held after the conclusion of each term in the work done in the several Lectures and Classes, open to all the pupils who may desire to take advantage of them. But none of these Examinations will be open to any person who has not attended the Lectures, or the Lectures and Classes, in the subject to which that Examination refers. Certificates will be granted to the Candidates who manifest sufficient merit in the Examinations.

The teacher to remain in the Lecture-room for some time after the conclusion of each Lecture and Class in order to answer questions, or solve the difficulties which have occurred to pupils, and to give advice as to the reading of text-books and other means of efficiently studying the subject.

Each Lecture to be accompanied at the discretion of the teacher by a syllabus distributed to the pupils, and by questions. Those who desire to answer these questions to do so in writing at home, and to be at liberty to submit their answers to the teacher for correction and comment.

The Class in each subject to be formed only from among those who attend the Lectures in that subject, and to consist of those who are desirous of studying it more fully. The Class, at the discretion of the teacher, either to take up the subject of the Lectures or cognate subjects bearing directly thereon and necessary for the better elucidation of the subject of the Lectures. The teaching in the Class to be more conversational than in the Lecture

The Towns to make all local arrangements; to provide Lecture-rooms; and jointly to pay to the University at the beginning of each of the two terms a sum of £375 for the supply of the teaching and Examinations; and also pay to the University a sufficient sum each term (estimated at about £*) to cover the expenses of travelling and printing incurred by the Lecturers on behalf of the scheme.[3]

The very first lectures given under the auspices of the Syndicate and the University of Cambridge, the first University Extension

* Blank in original.

46

courses ever to be given, were delivered at Derby and Leicester on the evening of Wednesday 8 October 1873. Since the first lecture at Derby began at 7.30 p.m. and that at Leicester at 8 p.m., Derby had priority. At Derby the arrangements were in the hands of a provisional committee of 38 – seven clergymen, eight members of the town council, three ladies, three magistrates and fifteen others.[4] The two principals were almost certainly T. W. Evans, the chairman, and the Rev. Walter Clark, the secretary. Evans was the local M.P. and a proprietor of the large cotton spinning mill at Darley Abbey, a few miles north of Derby, as well as various other commercial ventures.[5] Clark was headmaster of the local grammar school. He had obtained a Cambridge degree in classics in 1860 and went to Derby as headmaster in 1866. There he transformed it 'from the position of a third-rate grammar school, to that of one of the great public schools of the country'.[6] Clark and Evans were already involved in the provision of art and science classes at Derby under the South Kensington scheme. Evans was chairman of the local committee for art classes. These were held in the Derby Mechanics' Institute, but the Institute took no part in the organisation of the Cambridge lectures. An enquiry had been made on Stuart's behalf in April 1873, but a sub-committee appointed to consider his proposals recommended that nothing should be done 'until some definite arrangements are made by University officials'.* By the time that the Institute was reassured a provisional committee was already organising the courses and the sub-committee could only welcome 'so desirable an object'.

As a result of this failure there was a delay in completing the arrangements for Derby and, perhaps more seriously, the working classes were not directly represented on the organising committee. At Nottingham the mechanics' institute was much more active and at Leicester (where there was no longer an institute) the Working Men's College was closely associated with the first Cambridge courses. Derby was also unusual in deriving support for local lectures from those who opposed the Education Act of 1870. The local Conservative newspaper in welcoming Stuart's proposals used the occasion to support the voluntary principle in primary education.[7]

---

* Public Record Office, Ed. 29/26. Derby Mechanics' Institute minute books, 7 and 15 April 1873. The latter information was kindly supplied by Mr Chadwick of Manchester University who is working on a history of the Institute.

The first advertisements appeared in the Derby papers in mid-September:

There will be Two Terms, each lasting about twelve weeks.

In the First Term of Three Months, beginning October, Three Series of Lectures will be given as follows: –

1. A course of Lectures on 'Political Economy', by V. H. Stanton, Esq., M.A., Fellow of Trinity College, Cambridge. Adapted to Working Men.

2. A series of Lectures on some period of 'English Literature', by E. B. Birks, Esq., M.A., Fellow of Trinity College, Cambridge. Mainly for Ladies.

3. A course of Lectures on 'Force and Motion', by T. O. Harding, Esq., B.A., of Trinity College, Cambridge (Senior Wrangler, 1873). Mainly for Young Men engaged in commercial and professional pursuits.

SCALE OF FEES

| | | | |
|---|---|---|---|
| For Morning Lectures and Classes | 0 | 10 | 6 per Term |
| For all the Evening Lectures and Classes | 0 | 5 | 0 per Term |
| Ditto ditto for Artizans | 0 | 2 | 6 per Term |
| A Family Ticket, Transferable to any member of the same Family, giving admission to all the Lectures and Classes | 1 | 1 | 0 per Term[8] |

It was not until the following week that the times and dates of the courses were announced and that the meeting place would be the old grammar school building in St Peter's churchyard.* Even then it was necessary for Walter Clark to write to the papers explaining that, although the Literature course was mainly for ladies, gentlemen would not be excluded.[9] The demand for tickets was possibly small, because at the last moment it was announced that the first lecture would be free and held at the Guildhall. This advertisement did not appear until the day of T. O. Harding's lecture and the local papers noted that as a result the attendance was small.[10] Throughout the first year the arrangements at Derby were not as good as those at Leicester or Nottingham.

Harding's first lecture was chaired by T. W. Evans. Harding spoke principally of the value of education in science, but congratulated Derby on being 'amongst the first' in the establishment of extension courses. Birks' first lecture was at noon on Friday 10 October, with the Rev. Walter Clark in the chair. After the first week he lectured at noon on Wednesdays, and held his class at noon on Fridays. With some exceptions he lectured chiefly on the

---

* This building still stands and is now used as St Peter's Parochial Hall.

Romantic Poets. For Stanton's first lecture to the working men of Derby a chairman was not considered necessary.[11] Attendances at the later lectures were described as 'moderate' and it is clear that the Derby centre had a very uncertain beginning.

At Leicester the Cambridge courses had more competition from existing lectures and it was necessary to amend the programme as a result. In the absence of a mechanics' institute in the town several attempts had been made to provide lectures. The Literary and Philosophical Society had recently been reorganised. The Town Council had taken over complete financial responsibility for the Society's museum in 1871 and for the first time the Society was now able to pay professional lecturers.* From the autumn of 1872 the Museum Committee also ran a series of Saturday lectures 'for the adult classes'.[12] Both were only providing single lectures, but there were also two series of lectures in Leicester. The newly-appointed curator of the museum was organising science classes for South Kensington from his appointment in 1872.[13] A recently-formed Ladies' Lecture Society also provided courses in literary subjects from about the same date. In the winter of 1872 Henry Morley, Professor of Literature at University College, London, had lectured on English Literature and a second course was already planned for the winter of 1873. Stuart agreed that Birks' literature course should not be given at Leicester and that Morley's course should be substituted in the advertisements even though it was restricted to ladies and did not include classes or an examination.† An advertisement for all three courses appeared in the two local newspapers at the beginning of October and Morley's course was also advertised separately. Stanton's political economy lectures were on Wednesday evenings at the Town Museum in New Walk and the classes (for men only) at the Working Men's College in Union Street on Saturday evenings. Harding's lectures on Force and Motion were at the Museum on Monday evenings and his class was there on Thursday evenings. Morley's ten lectures

* F. B. Lott, *Leicester Literary and Philosophical Society* (Leicester, 1935), p. 65. Although the Society presented its museum to the town in 1849 it had continued to run it and pay the curator.

† *Leicester Chronicle*, 4 October 1873. Ellis, *Nineteenth Century Leicester*, p. 287. This special arrangement had been forgotten as early as 1880 when the *Calendar of Cambridge Local Lectures* was compiled. On p. 61 it is stated that Birks lectured at Leicester. This mistake has been repeated several times since – e.g. J. Simmons, *Life in Victorian Leicester* (Leicester, 1971), p. 69.

were on Friday mornings at 11.* The Rev. T. J. Vaughan wrote to both papers as chairman of the local committee for university extension recommending the courses and explaining that text-books could be obtained from the museum curator who was acting as local secretary. With his letter he also sent letters from Stanton and Harding outlining their courses in the same way that James Stuart had done for the workmen of Crewe. The cost of the two Cambridge courses of lectures and classes was five shillings a term, but the Working Men's College would only admit men to Stanton's classes, so that it was possible to pay half a crown for the lectures only. Morley charged half a guinea for his ten lectures or two shillings each, but the local committee had arranged for a two guinea ticket which enabled any member of a household to attend any of the three courses.[14]

The first lecture at Leicester was not reported by the local newspapers. Stanton probably gave an introductory talk, like Harding's at Derby, on 8 October because both papers printed a summary of his second lecture.[15] Brief notices of Morley's and Harding's first lectures appeared in the *Chronicle*. The former was attended by 150 ladies and the latter by 40 students. Harding's platform, however, was graced by three local clergymen and an alderman.[16]

The arrangements at Nottingham were different from both Derby and Leicester. An energetic local committee issued a poster headed 'University Education for the People' listing the programme for both the Michaelmas and Lent terms.[17] Free inaugural lectures were given in the lecture hall of the Mechanics' Institution on 9 and 10 October. The rest of Birks' lectures and classes were at the School of Art and Harding's in the Lecture Hall in Belgrave Square. The fees for English Literature or Force and Motion were five shillings a term for lectures and five shillings a term for classes. 'Persons engaged in Tuition' were admitted half price and two or more members of the same family for 7s 6d each. For Political Economy the lectures were 1s 6d a term and the classes 2s 6d. Twenty-two local booksellers and the local newspaper offices had been persuaded to sell tickets.

At Nottingham the local Mechanics' Institution took the

---

* *Leicester Chronicle*, 4 October, and *Leicester Journal*, 3 October 1873. The Working Men's College no longer stands in Union St, but the lecture room in the basement of the City Museum can still be traced.

initiative in organising the first courses and it was supported by the local Trades Unions Council. The Institution took the initiative because it felt that the Education Act of 1870 had altered the role of mechanics' institutes which ought in future to concentrate on the higher education of the working man. As early as 1871 the Nottingham Institution had passed a resolution at its annual general meeting calling for instruction 'in those subjects which are most important to them as workmen, as fathers of families, and as sharers in the political power of the country'.* A shortage of competent lecturers at Nottingham prevented anything being done until James Stuart made his offer. The Institution provided help and the use of their lecture rooms, but it did not directly sponsor the courses because it was found 'that a sum of money, as well as considerable time and labour would be required'.[18] Local trade unions also gave financial aid and an independent committee was formed which included the Anglican Canon Morse, the Congregational minister J. B. Paton, Richard Enfield to represent the Institution,† and W. Hicking from the Trades Council. A determined effort was made at Nottingham to appeal to working men and a poster issued by the committee ended with this plea:

Now let the Working Men of Nottingham give their heartiest support to a movement which was begun and has been carried out largely for their benefit, that they may receive the blessings of that 'higher Education' which the University can give them. Let them attend these Classes in such numbers, as to show the University and the Country that they know the value of Education, and gladly seize the opportunity that is now open to them.[19]

The first lectures were not held at Nottingham until 9 October because the Goose Fair, Nottingham's annual celebration, did not end until the previous evening. The organisers of Local Lectures realised from the start the importance of avoiding such local events. In 1875 the Rev. W. Moore Ede said that it was 'useless to begin lectures until the Nottingham Goose Fair and the Leicester Races were over'.[20] On Thursday morning E. B. Birks

* 33rd Annual Report of the Institution (1871), p. 14. Nottingham Daily Guardian, 31 January 1871. Nottingham Journal, 10 May 1873. The final phrase of the quotation refers to the recent extension of the franchise by Disraeli.
† Enfield was a local lawyer and a member of the Unitarian church at High Pavement. See J. C. Warren, Catalogue of Portraits, etc. at High Pavement Chapel (n.d.), p. 24.

spoke to a 'large audience chiefly of ladies', and in the evening V. H. Stanton had a 'crowded meeting'. Both spent much of their time emphasising the importance of their subject, but Stanton also included general remarks on the two year experiment which Cambridge had begun. On the following evening T. O. Harding gave a similar lecture to a 'very large and enthusiastic meeting'.[21]

One of the Nottingham organisers was able to persuade *The Times* to comment editorially on university extension. On 9 October 'University Education in the Provinces' gave a favourable account of the growth of Owens College at Manchester and 'the lectures at Nottingham'. The Nottingham newspapers reprinted this with considerable pleasure.[22] It was this early publicity, together with the temporary cessation of lectures at Derby and Leicester a few years later, which has led to claims that Nottingham took the initiative in organising lectures.[23] It is true that at the meeting in Nottingham on 10 May there had been wild proposals that Nottingham should obtain the best lecturers from both Oxford and Cambridge for its courses,[24] but in the end they took what James Stuart had to offer just as Derby and Leicester did. The first three Cambridge centres were deliberately chosen by James Stuart.

His choice was not as eccentric as it might seem at first. In the northern industrial towns where he had previously given courses, much was already being attempted. The North of England Council was already providing courses for women in the major towns,[25] and there was some hope that the mechanics' institutes and co-operative societies might do the same for men. The larger populations of these towns not only made the organisation of such courses easier, but also encouraged free-lance lecturers in some subjects. The population of the three Midland towns was much smaller (about 70 000) and each was just emerging from a period of difficulty caused by the failure of the framework knitting trade. At Derby the establishment of the Midland Railway headquarters in 1844 had brought prosperity. At Leicester boot and shoe manufacture and other trades had diversified the town's industry. At Nottingham new machinery for making hosiery and lace had helped to solve its problems. In each town there was sufficient prosperity by 1873 to stimulate the demand for further education, but little idea of how it could be supplied. Single lectures, science and art classes or ladies' lecture societies were clearly insufficient.

When the Cambridge scheme was first proposed in Leicester, the editor of the local Liberal paper wrote:

In Leicester, as in other towns, Mechanics Institutes once did much to promote a love of mental improvement among young men of the middle and working classes. Unfortunately, however, for various reasons, those institutions gradually collapsed, and others have not been established in their place. In consequence Leicester, and other large towns, are now in a worse position in this respect, than they were thirty or forty years ago.[26]

At Nottingham the Mechanics' Institution had been hoping to organise courses of lectures for two years without success.[27] Therefore it was easier to prove to the University that there was a need for university extension in these towns than it would have been in Birmingham, Leeds or Liverpool. Stuart may also have thought that there was more chance of success where there were fewer rival lectures.

It was also Stuart's intention to combine the centres into circuits which would share the cost of the lecturers' travelling and accommodation, and the three Midland towns were well situated to form the first circuit so long as they would agree to the same courses. Less than twenty miles apart, each was linked to the others by the Midland Railway Company's main lines from Trent Junction. They were also the three nearest industrial towns to Cambridge, and Leicester had good rail connections with the University through Kettering or Peterborough. James Stuart said later that 'we started on too ambitious a scale, and we had to suffer for it',[28] but the basic soundness of his choice can be judged from the later career of the three centres. Nottingham remained a Cambridge centre without a break until it was handed over to its own University College (itself a Cambridge creation) in 1914. Leicester, with one brief interval, was a Cambridge centre until 1931 when its University College took over the work. Derby has the distinction of being the last Cambridge centre outside East Anglia.

The beginning of Lent term 1874 saw the opening in Yorkshire of four new centres. In December James Stuart had visited Leeds, Bradford, Halifax and Keighley to address public meetings. At Bradford and Halifax the meetings were sponsored by the local mechanics' institutes. The Bradford meeting on 9 December attracted only 'a very meagre attendance' and only a general resolution approving of university extension was passed.[29] Two evenings later at the annual meeting of the Halifax Institute there was much more enthusiasm. £230 had already been raised, and the Mechanics' Institute, the Literary and Philosophical Society and the Co-operative Society had all promised to help in different ways.[30] However the centre of all the activity in the West Riding was Leeds where the Ladies' Educational Association had taken the initiative as early as June 1873. Its committee had invited 'a few gentlemen interested in the matter to confer with them as to its practicality'.[31] The Association was already organising courses by some of the Cambridge lecturers, but it was anxious to extend the benefits to others. Members of the committee approached the Co-operative Association, the Trades Council and the Early Closing Association in Leeds as well as individuals in Bradford, York, Keighley, Huddersfield, Dewsbury, Halifax and Sheffield. On 25 July they organised a conference at which a joint committee representing Leeds, Bradford, Halifax and Keighley was formed.[32] In Leeds the Ladies' Association guaranteed £100 in tickets for the first term. Because the Association sponsored a history course by Professor Seeley in the autumn, it was decided to begin university extension in the West Riding in Lent term. The courses were opened with a grand inaugural meeting at the beginning of January. Edmund Baines, a local M.P. and President of the Local Government Board, was in the chair; Josephine Butler and her husband, James Stuart, and representatives of the Inspectors of Schools, the Leeds School Board, the Leeds Co-operative Society and the Leeds Trades Guild of Learning were on the platform.[33] There were courses on Political Economy, Constitutional History and Physical Geography. The Rev. W. Moore Ede of St John's lectured at Halifax and Keighley on political economy, and the Rev. William Cunningham lectured on the same subject at Leeds and Bradford.[34] They took the place of V. H. Stanton who had previously lectured at Leeds and was now kept in Cambridge by increased college responsibilities. Both his substitutes were so successful that in the following year Cunningham was appointed organising lecturer in South Lancashire and Moore Ede in the Midlands.*

It was at this that the first independent attempt was made to promote university extension. Although Birmingham had produced

* The duties of the post of organising lecturer were very similar to those of resident tutor at the present time.

one of the memorials to Cambridge in 1872, it never became a Cambridge centre. It was apparently because of local feeling that the town established an Association for Promoting Higher Education in Birmingham at a public meeting on 17 December 1873. The meeting was chaired by Joseph Chamberlain and attended only by local people.[35] It was so successful that in the following winter it ran 41 courses, but many of these were not of university standard.[36] Almost all the lecturers were local, but the students were entered for the Cambridge Higher Local Examinations.

There are no accurate records of the success of the first year of the Cambridge local lectures. The Syndicate did not know how many attended the lectures or the classes and no lists of examination results were published. James Stuart did not attempt to collect statistics until 1875. Local newspaper accounts are usually vague and probably unreliable. At Derby 'the first term of these lectures was not so successful as could have been hoped, the attendance having been rather small'.* At Leicester the chairman of the Museum Committee said that: 'these lectures did not attract large audiences, probably because the lecturers have not remembered that the general adult public do not at present feel a yearning desire for dry knowledge, although that desire may be kindled by putting the knowledge before them in a lively and interesting manner'.[37]

However this writer was probably concerned about the effects of the Cambridge courses on the single lectures sponsored by the Museum Committee on similar subjects. The course at the Working Men's College (from which women were excluded) was said to have been 'very fairly attended'.[38] For Nottingham there are detailed figures of the tickets sold:

|  | Lectures | Classes |
|---|---|---|
| *Michaelmas* | | |
| English Literature | 378 | 194 |
| Force and Motion | 108 | 70 |
| Political Economy | 316 | 156 |
| *Lent* | | |
| Physical Geography | 228 | 101 |
| Astronomy | 92 | 30 |
| Constitutional History | 117 | 42[39] |

* *Derby and Chesterfield Reporter*, 16 January 1874 (BEMS 1/6, 12). The *Derby Mercury* (14 October 1874) said the attendance was between 30 and 50.

These are the most accurate figures which we have. They are very similar to later figures of tickets sold at Nottingham,[40] but they are not comparable to the attendance figures from Leicester and Derby.

No courses were given in Easter term 1874, because it was unlikely that they would be well-attended. The North of England Council had realised this a few years previously,[41] and it has been a constant principle in university extension ever since. But plans were made to open new centres in South Lancashire and the Midlands in October 1874. At Liverpool a large public meeting had been held as early as January.[42] It was chiefly noteworthy for the appearance on the same platform of Miss Davies and Miss Clough which led to a local revival of the controversy between them about which Cambridge examination was more suitable for women. William Cunningham was sent to Liverpool as organising lecturer in the summer and soon planned an ambitious programme in collaboration with the local committee. There was already an active Ladies' Educational Society (to which Stuart had lectured) and Societies for the Promotion of Higher Education were formed at Liverpool, Everton and Waterloo.[43] There were several different committees in Liverpool and others at Birkenhead and Southport. In the Michaelmas term Cunningham lectured on Political Economy 'to persons of the working class', and on Economic Science 'to persons of better education'. There were also courses on Physical Geography. On the administrative side Cunningham's mother noted 'how he kept up unfailing good humour while concerting various arrangements with the Secretaries of small Associations, and in putting right the blunders and difficulties which are sure to arise, where many different parties are concerned'.[44] By his efforts other local centres were begun in the area in the next few years and a contribution made to the establishment of the new University College.*

By this time a pattern for the establishment of new centres had developed. If possible a local sponsoring body was found – the Literary and Philosophical Society or the Mechanics' Institute. A

---

* The important part which the Association for the Promotion of Higher Education in Liverpool and the Cambridge courses which it organised played in the establishment of the University College in 1878 has been forgotten locally. See T. Kelly, *Adult Education in Liverpool* (Liverpool, 1960), pp. 33–5; *University Recorder* no. 54 (1970), pp. 10 and 11; University Archives, Prof. Ormerod's History of the University.

public meeting was called (often a town meeting presided over by the mayor) at which James Stuart or one of his colleagues explained the Syndicate's work. Local newspapers were persuaded to print paragraphs about extension courses in other towns and one of the Syndicate's leaflets was distributed. If successful a local committee was appointed by the public meeting and it made all the arrangements with Cambridge and established a local guarantee fund. The local committee usually put advertisements in the local papers and about a week before the lectures began they would be asked to print letters from the lecturers giving advice about books and background knowledge. By December 1875 the Syndicate was able to condense much of this information into a leaflet of advice 'for those contemplating the establishment of courses of teaching'.[45] By James Stuart's efforts the work expanded rapidly. In October 1874 new centres were also established in the Potteries and Sheffield; in January 1875 four new centres were opened and the following October there were thirty centres in existence.* All the organisation was done by James Stuart and one clerk and, although he never lectured for the Syndicate, he travelled the country describing university extension and urging Oxford and the larger towns to take up the work. He well deserved the Syndicate's tribute: 'To Mr Stuart's activity, judgement, and power of organisation, the Syndicate consider the success which has hitherto attended the establishment of these local lectures is mainly due'.[46]

At the end of Easter term 1875 the Syndicate's powers ceased. Although it seemed unlikely that the work would not be allowed to continue, every effort was made to publicise the work which had been done. The Syndicate instructed James Stuart to prepare a draft report on its activities in December 1874.[47] The report was also to ask 'the Senate to appoint a managing Syndicate'. He also took the opportunity to comment in detail on the financial arrangements.[48] From the figures for attendance during Michaelmas term he concluded that the University provided the courses at a cost of nine shillings a head, but this figure did not include the local costs of hiring a room or advertising. However he was able to point out that the Syndicate insisted on all expenses being met by the local committee so that there was no charge on the

* These figures are taken from the Syndicate's *Reports* (BEMS 22/1, 58–64) and do not completely agree with the other information available (e.g. *Calendar of Cambridge Local Lectures, 1875–80*).

University. The report was accepted by the Council on 8 March 1875, and the Senate was asked to appoint a Managing Syndicate 'consisting of the Vice-Chancellor and twelve other members of the Senate, elected by Grace, four of whom shall retire in rotation on the 20th November in every year'. Its duties were:

> to organise and superintend courses of Lectures and Classes at such places as they may select, to hold such Examinations as they may deem expedient and to fix the times and places of the same, and the fees to be paid by all persons who may be admitted thereto, and to determine generally all other matters connected with the said Lectures and Classes and Examinations, on condition that all the funds which may be necessary are guaranteed by the local authorities.[49]

The Senate unanimously approved this on 18 March.

In February the old Syndicate had also instructed Moore Ede to prepare a report on the work in the Midlands and the Vice-Chancellor ordered this to be printed on 1 March.[50] Moore Ede began by setting out the usual objections to local lectures – that their effect was ephemeral and that they did no real good to students. He answered them conclusively:

> In order to answer this question it is necessary to ascertain whether the lectures and classes are really valued by the students, and whether they meet a want unsupplied by other educational agencies.
>
> In answer to the first point we have the following proofs:
>
> (a) The large number of tickets sold for the courses of lectures; this shows a large increase since last year in the case of Derby and Leicester and no real diminution in the case of Nottingham.
>
> (b) The very regular attendance of the students even in the very worst weather, and the very slight falling off in numbers towards the close of the term; and, when it is borne in mind what a large proportion of the students are busily engaged in offices or workshops the whole day, it is no bad criterion of the value of this means of education to them, that we find them after a severe day's toil turning out week after week to sit for an hour or two to listen to lectures on some branch of study in which they happen to be interested.
>
> (c) Everywhere the old students have been most desirous of the continuation of the scheme, and have done their best to induce their friends to join the lectures. At Leicester, for example, when the Committee had decided to discontinue the lectures, the old students met together and resolved to make an effort to obtain the reversal of the Committee's decision.
>
> (d) Many have expressed the immense benefit they have derived from the instruction received at the University Lectures; one wrote to the effect that he would rather 'have a term at Cambridge than a princely

fortune'; another, a lady, said she, 'had never in any study learnt the art of thinking so well'; another, a working man, remarked, 'Don't it open a fellow's head just?' and that expression of his conveyed the feeling of many of the students. Indeed the evening lectures do open up to the audience many new lines of thought, and above all teach them the difference between thorough knowledge and a mere smattering. At Keighley in Yorkshire a number of students used to meet together once a week, and, aided by their syllabuses, discuss the last lecture together, and then bring to the lecturer next week any difficulties they found themselves unable to explain. The same plan was pursued by the Leicester students ...

(e) Everywhere the value of the lectures for ladies is recognised, schools gladly bring their more advanced pupils to the lectures and classes, and some parents who have sent their children to other towns for education, talk of bringing them home, in order that they may have the advantage of the thorough teaching which the University Extension gives.

(f) Another sign of the value the students set upon the scheme, is to be found in the trouble they take to answer the questions set. Week after week many of them devote much of their leisure time to writing their answers, and, when one looks over the frequently ill-spelt sheet, the formation of the letters tells that the hands which have guided the pens are but little accustomed to such a task, and it becomes difficult to mark as wrong that which one knows has cost so much toil and trouble. A further proof is to be found in the fact that the Nottingham Trades Unions have subscribed to the support of the lectures and also become Guarantors. In Keighley the Co-operative Society purchased a number of tickets which were balloted for among its members.[51]

Moore Ede also maintained that the local lectures did not duplicate work being done elsewhere by taking each possible alternative in turn:

(1) Mechanics' Institutes, though they have done good service in education, and are still by their libraries and reading rooms furthering the dissemination of useful knowledge, yet do not provide for the higher systematic instruction of their members. The various lectures, which are given in their rooms, arouse interest in the minds of the audience and amuse them for an evening, but no solid teaching is afterwards provided for those in whose minds interest has been excited. The same is true of all other popular lectures, such, for example, as the Saturday Free Lectures at Leicester.* The University Scheme, however, does much more than this ...

(2) Working Men's Colleges do a very good work, but they generally give instruction in the more elementary branches of knowledge, and not anything which can be called higher education. At Halifax one of our lecturers gave a course of lectures at the Haley Hill Working Men's College, and I was myself requested to do the same if I returned to

* Organised by the Museum Committee at the Town Museum.

Yorkshire; thus it is plain that, so far from our coming in conflict with these Colleges, they are glad to avail themselves of our assistance. Many of my class at Halifax were also students of the Working Men's College ...

(3) The Government Science and Art Classes, which perform a great work in the education of the masses, do at first appear to come into collision with our scheme, but in practice we find that there is plenty of work for both, and that the two schemes work very well side by side, and indeed play into one another's hand, if I may use the expression. The Science teachers ... generally confine themselves to teaching the more elementary branches of science, and their tuition is somewhat similar to that given to the upper classes of schools ...

(4) There remain the Local Colleges such as Owen's College, Manchester. As Leeds is the only town where the University lectures are held in which there is any College of this sort, these Local Colleges cannot be said to be doing the work which the University proposes to do. The Yorkshire College of Science [at Leeds] is of too recent foundation to say whether it will do our work or not. At any rate these Local Colleges are but few in number, and wherever it is found that any such institution is really educating all classes of society it will be an easy matter for us to withdraw, for there will always be plenty of towns where our services will be required.[52]

Although the Cambridge plan of courses of ten or twelve lectures had considerable advantages over the single lectures of mechanics' institutes, Moore Ede considered that the courses should be arranged to provide a continuous course of instruction over three years (six terms) which would be recognised by the University. He suggested that students who successfully completed this should be given the degree of A.C. (Associate of Cambridge) and exempted from one year's residence at Cambridge if they graduated there. He hoped that this would not only benefit the students, but would provide more stable financial backing for courses. He sketched out a possible three year curriculum:

First term – Logic      Second term – Constitutional History
Third & fourth terms – Political Economy
Fifth term – Social History      Sixth term – Political Philosophy.[53]

Some of these ideas were afterwards adopted in the affiliation scheme of the Syndicate.*

It is not clear from the surviving correspondence whether Moore Ede had discussed his report beforehand with James Stuart. Some of the proposals had been suggested in Stuart's

* See below, p. 138. In 1898 Moore Ede repeated much of his report in an address to the Cambridge Conference (*Report of a Conference 1898*, pp. 18–23).

address on University Extension, but the details were probably Ede's own. Stuart certainly adopted them enthusiastically, urging the Vice-Chancellor to have the report printed and circulated and suggesting that a revised version might appear in one of the London literary magazines.[54] He also had the proposals discussed at a conference of local centres* at which he planned to consider the future progress of university extension work.

Early in March 1875 a duplicated letter from James Stuart invited the centres to send their secretaries and as many other members as possible to attend a conference in the Cutlers' Hall at Sheffield at noon on 31 March.† Sheffield was chosen for the meeting because John Moss, one of the secretaries there, had volunteered to make all the arrangements.[55] Stuart suggested two topics for discussion:

1. Measures which tend to promote the administrative efficiency of the work, the attracting of large numbers of students, and the enlisting of public interest and support.

2. Whether any and what further development [*sic*] of the teaching is yet desirable and practicable, such as the introduction of a more methodical curriculum, the establishment of elementary classes preparatory to the present teaching,‡ and advanced courses for second year students.

But he also asked for other subjects to be sent to him in advance, and these were incorporated into a printed sheet given to all the delegates:

1. What are the best methods of making known to the greatest number of people the nature and advantages of the teaching and of securing a large attendance at the Lectures of (i) young men of the middle classes, (ii) young women, (iii) the working classes ?

2. – CLASSES
Is it advisable that the University should adopt any modification of the system of classes in connection with the Lectures, or of the conditions of admission to the classes ?

3. – THE PECUNIARY ASPECT OF THE SCHEME
How far are the expenses to be met by fees alone ? What modifications in the system of fees are desirable ? *e.g.* should the fees charged in any place be uniform, or should a higher fee be charged than at present and

---

* Because no report of its proceedings was ever issued this first conference was completely forgotten in later years – even by James Stuart.

† The draft for the letter is in BEMS 37/1, 342. The surviving lithographed copy (BEMS 1/4) is dated 6 March, but it was issued earlier (see BEMS 37/1, 62, 67).

‡ A few courses of this kind were held at Nottingham and elsewhere in the early years of the Syndicate.

remissions made under certain circumstances, or should the fees be different for different classes of society ?

4. – MANAGEMENT
   (i) Committees.
   (ii) The formation of sub-committees.
   (iii) The formation of federal committees.
   (iv) Arrangements between the University and groups of towns.

5. – ADAPTATION
Can any further steps be taken for adapting the teaching to pupils in various stages of advancement ?

Preparatory Classes
Advanced Classes

6. – METHODIZATION
   (i) The question of a curriculum of study extending over several years.
   (ii) The pecuniary aspect of a curriculum.
   (iii) In drawing up a curriculum, should morning and evening courses be considered separately ?
   (iv) Its application to small and to large towns.
   (v) What would be the effect of a degree or final certificate given to a pupil who has passed successfully through such a curriculum ?
   (vi) What assistance can be rendered by local teachers in the carrying out of a more complete scheme ?

7. – ADDITIONAL QUESTIONS ON WHICH INFORMATION IS DESIRED.
   (i) Should there be free inaugural lectures ?
   (ii) Should there be admissions to single lectures ?
   (iii) Should the courses be of three or of six months ?
   (iv) How many lecturers is it most advantageous to employ at the same time in a town ?
   (v) Is it advisable to repeat the courses in different parts of the town ?*
   (vi) What are the best dates for beginning in the autumn and after Christmas respectively ?
   (vii) Is it possible to have a uniform time fixed for the examinations ?
   (viii) What is the best arrangement of a lecturer's work ?
   (ix) Should the same lecturers continue in a town for several terms in succession ?
   (x) Is it advisable to locate a lecturer entirely in one town, so that he may take morning and evening courses ?[56]

This document summarises all the problems and difficulties encountered during the first two years of university extension and it is a pity that no shorthand record of the discussion on them was kept. It would have been very informative.

* This was being tried at Liverpool.

A copy of Stuart's introductory remarks to the conference has survived. He made two definite assertions in this – 'We have proved that there is a need and that it can be supplied ... I make bold to say that the pecuniary difficulty has been removed.'[57] There is also a note asking for the draft of a resolution about the 'curriculum'. A pencil draft asking in guarded terms for the University to recognise successful university extension students is apparently the reply to this.[58] Another, more coherent, draft is concerned with the burden of work on James Stuart, who had recently been ill.

In the opinion of this Conference the movement has now assumed such proportions that it is exceedingly important to provide for the lessening of the work hitherto almost exclusively entailed upon the central authority: either by the appointment of special local secretaries for groups of towns, the formation of federal Committees or in some similar manner.

Further this meeting desires [most] respectfully to record its marked sense of the urgent necessity for the appointment of some thoroughly competent gentleman at an adequate salary as Mr Stuart's coadjutor to share with him the rapidly increasing work devolving upon the central authority [and being well aware of the liberal views of the Syndicate as shown in their recent meeting in the Arts School hope that this will receive their early consideration].*

The new Syndicate took heed of the last resolution at its first meeting on May 15 by appointing V. H. Stanton as joint secretary with Stuart.† Vincent Henry Stanton was a few years younger than James Stuart; the son of an Anglican clergyman who had previously been chaplain of Hong Kong. Stanton won a minor scholarship at Trinity in 1866 and was elected a fellow in 1870. He was a member of the Grote Club and one of the Trinity reformers.[59] After his ordination he became one of the first lecturers for the Syndicate. Even after increased work in the University prevented him from taking courses he maintained his interest in the work of the Syndicate and lectured at its Summer Meetings. At Trinity he was successively dean and tutor: in the University he was lecturer and then professor of divinity.

The Syndicate also considered a proposal to issue certificates to

* BEMS 37/1, 347. The words contained in square brackets were added later.
† Although Stanton's name is omitted from the list of secretaries published by the Syndicate in 1923 (*Historical List of Secretaries and Lecturers*), the minutes (BEMS 4/1, 24), Stuart's own *Reminiscences* (p. 173) and Browne's *Recollections* (p. 128) are all equally certain about his appointment.

students who were successful in the examinations. On 2 December after various unreported debates certain proposals were 'adopted in principle',[60] and after further discussion appeared as an appendix to the Syndicate's report of 12 June 1876:

1. At the end of each course of lectures and classes Certificates of Attendance and Examination will be given to all who, having attended the lectures and classes, also pass the examination successfully. The Certificate, which will be signed by the lecturer under the authority of the Syndicate, will specify the subject and duration of the course and the distinction, if any, which the Student has gained in the examination, and in the class-work during the course.

2. Any Student who obtains Lecturer's Certificates of Attendance and Examination for courses extending over a period of at least six terms in subjects all of which belong to the same group will, on presentation to the Syndicate of these Certificates, become entitled to receive a Certificate signed by the Vice-Chancellor of the University, stating that the Student has passed through a course of study approved by the Syndicate, and signifying the courses attended and the distinctions, if any, which have been gained.

3. The groups referred to in the preceding paragraph shall be, (1) Literature, Language and History; (2) Political Economy, Logic, History and Moral Science; (3) Natural Science. The Syndicate intend, as soon as adequate arrangements can be made, to prescribe a definite course of study in each of these groups.

4. The Syndicate reserve to themselves the right of altering these regulations from time to time or of rescinding them, and of requiring satisfactory evidence of a sufficient elementary knowledge on the part of the pupils, and also of charging such a capitation fee as they may from time to time deem advisable.[61]

Probably because the regulations contained no mention of granting exemption from the University's Previous Examination or of conferring a Cambridge degree, the Senate raised no objections to them, and the Syndicate was able to issue the lecturers' certificates at once. The first Vice-Chancellor's Certificate was issued in 1879 – after the regulations had been further amended.[62]

With the increase in the number of centres and the appointment of joint secretaries it became necessary to have an office. A room was rented at 3 Trinity Street (which was then a lodging house and offices) almost opposite Senate House.* A clerk, W. Ridgeway, was installed there. Ridgeway was a Sheffield man – probably recommended by one of the local committee – and he returned

---

* The house still exists. It is next to Bowes and Bowes' bookshop in Trinity St.

there at the end of January 1876 because of homesickness and domestic problems.[63] He was succeeded by a Mr Stanley about whom nothing is known. The clerk's duties were to answer routine correspondence, despatch parcels of syllabuses and examination papers, prepare the certificates and collect statistics.[64] For the first time it became possible to gather figures of attendance and to record the names of those successful in the examinations. Both Ridgeway and Stanley were paid about 13s a week, and the office rent was £22 10s a year. This sum of about £50 was paid by Stuart in 1874–5, but it was charged to the Syndicate in the following year.[65]

This payment made out of his own pocket, but not recorded as a donation to the work is typical of the early Syndicate accounts. Just as he was to do later with the Engineering Workshop accounts, Stuart kept only the minimum information and kept the Syndicate solvent with his donations. For the first year there is a tentative balance sheet in the minute book.[66] According to this £1225 was received from local centres, £1149 was paid to lecturers and £48 to examiners, while £18 16s 6d was used for printing. The balance in hand of £9 3s 6d went to Stuart 'for helping to pay for certain clerks work'. For the second year (1874–5) there is another tentative balance sheet and Stuart's pocket notebook.[67] The latter contains payments such as the £48 for 'Office Expenses' which do not appear in the former. According to the balance sheet the receipts were £2885 10s and payments £2878 6s leaving an apparently favourable balance of £7 4s for the following year, if Stuart's donations are ignored. For the third year (1875–6) and the last for which Stuart was responsible there is another balance sheet, a few notes in his pocket book and an account book.[68] The account book was intended to be a ledger and a number of pages were assigned to each centre and each lecturer. For each the amount due is recorded and then a note of each payment made. Unfortunately it was not kept up to date after Ridgeway left in January 1876 and so cannot be compared with the balance sheet. The latter showed an apparent surplus of £52 19s 4d, but concealed the fact that Barnsley had not paid the £67 6s 11d included in the receipts. There was a loss of at least £14 7s 7d.

It was unfortunate that at this point when the work was expanding rapidly and the University should have been pressed to take some financial responsibility that both secretaries were obliged to resign. James Stuart had been professor of mechanism for a year and a severe illness made it important that he should give up some of his responsibilities. V. H. Stanton was appointed examining chaplain to the Bishop of Ely in 1875 and senior dean of Trinity in 1876. Although both were still very interested in university extension and did as much as they could to help as members of the Syndicate for many years to come, neither was able to direct the work in a way which was needed.

# 5

# G. F. Browne and R. D. Roberts, 1876-91

The new Secretary of the Syndicate was the Rev. G. F. Browne – a man much more typical of the average Cambridge don at this time than either Stuart or Stanton. He was an Anglican clergyman and belonged to a clerical family, and he was a Conservative and deeply involved in University politics. He was appointed Secretary chiefly because he was already secretary of the Local Examinations Syndicate, but there was probably also a feeling that he was a 'sound' man. It has frequently been said that his influence was good because he brought the expansion of local lectures work to an end and 'consolidated' the activities. During Browne's secretaryship Stuart took this view in public, but some of his other remarks contradict it. In fact the sharp reduction in the number of courses which Browne made and his abandonment of missionary activities was little less than tragic. Retrenchment could never solve the Syndicate's financial problem, and reliance on a subsidy from Local Examinations only disguised its seriousness.

However, Browne's influence was to some extent counterbalanced; at first by the voluntary work which James Stuart continued to do, and later by the enthusiasm of R. D. Roberts who was assistant lecture secretary from 1881 to 1890. Stuart remained a member of the Syndicate and for some years was particularly concerned with its relations with the local colleges in the provinces. He had always envisaged university extension as including the establishment of university colleges – embryo universities in the larger industrial towns,[1] and the Syndicate helped almost all of them at different times in their development and in different ways. At Birmingham and Manchester the Syndicate organised some courses in conjunction with Mason's College and Owen's College. At Newcastle upon Tyne the Syndicate's courses had to be arranged through Durham Univer-

[1] For References see p. 200.

3-2

sity, but Moore Ede and at least one other Syndicate lecturer later taught at Newcastle.* At the other university colleges the assistance which the Syndicate rendered varied from courses which supplemented local efforts in the field, as at Liverpool, to the founding of a university extension college at Nottingham. Leeds and Sheffield fell between these two categories. At Leeds at first the local lectures were run independently of the Yorkshire College of Science. The local committee felt that: 'not only was the College of Science intended more directly for those seeking technical rather than higher education, but also . . . it was as yet unknown whether its benefits would extend to all the classes whom it was hoped University Extension would influence'.[2]

Therefore until 1877 the Cambridge courses were confined to non-science subjects, while the College did nothing on the arts side. When in 1877 the College dropped Science from its title and appointed professors of history and literature, the Syndicate withdrew from the town by agreement.[3]

At Sheffield the Syndicate began work before there was a local college. The inaugural meeting there on 10 December 1874, at which Stuart spoke, was chaired by the mayor, Alderman Mark Firth.† Firth, the owner of a local steel company, was personally interested, and 'before the close of the second session' he 'promised to provide a suitable building which should be specially devoted to purposes connected with the moral, intellectual and social advancement of his fellow townsmen'.[4] In the autumn of 1879 Firth College was opened in the centre of the town with two lecture halls, a library and a few offices. On 5 June in the previous year the local Extension Committee had asked the Syndicate to affiliate the College to Cambridge University,[5] but there was a long delay before the necessary powers could be obtained and Firth College was obliged to go ahead independently.‡ Nevertheless between 1887 and 1889 the Syndicate organised three courses at Sheffield,

* The two colleges at Newcastle were part of the University of Durham. However the request to the Syndicate for affiliation there came jointly from the Literary and Philosophical Society and the Science College (*Report of a Conference*, 1887, p. 57 – BEMS 28/1). See E. M. Bettenson, *The University of Newcastle upon Tyne* (Newcastle, 1971), pp. 24 and 88.

† *Sheffield Independent*, 4 and 11 December 1874 (BEMS 1/6, 26 and 27). There had been a preliminary meeting on 3 December.

‡ The necessary statute had to receive the approval first of the Cambridge University Commission and then of Parliament. It was not in force until late in 1881 (*Eighth Annual Report* – BEMS 22/1, f. 74).

and from 1898 to 1905 English Literature courses for pupil teachers jointly with the College.[6] Meanwhile in 1897 Firth College, the Sheffield Medical School and the local technical school had been combined to form University College which in 1905 became the University of Sheffield.

The Syndicate was also closely associated with the foundation of Liverpool University. From the time of Stuart's first lecture at Liverpool in October 1867 the Liverpool Ladies' Educational Society had taken courses from Cambridge lecturers, and by 1875 the Syndicate's lecturers gave courses for ladies at the Royal Institution.* Meanwhile it had been decided to offer Cambridge courses to other classes. On 20 January 1874 a public meeting at the Town Hall had led to the establishment of a Society for the Promotion of Higher Education in Liverpool,[7] and the inaugural Cambridge lecture under its auspices was given at the Cotton Brokers' Association Rooms on 5 October 1874.† The courses were given in the Temperance Hall to 'artisans' and in the Medical Hall 'to persons of a better education'.[8] A very active local committee started courses in the neighbourhood of Liverpool – at Birkenhead, Bowdon and Southport during the first term, and later at Chester, Everton and Waterloo.[9] They were aided in this by the appointment of an organising lecturer.

Despite the competition from courses of free lectures at the Public Library which had begun about 1865 and an annual deficit met by subscribers, the Society was very successful and in 1878 the Library Committee also applied to the Syndicate for courses. Moore Ede gave the first course of twelve lectures on Political Economy at the Library beginning on 9 October 1878. The corporation paid all the expenses and advertised that 'It is not intended to make any charge'.‡ Several more courses were provided on this basis, but in 1880 the Syndicate suggested a course on the history of Great Britain as illustrated by archaeology. The

---

* 'The Lectures of the Ladies' Educational Society are given two mornings every week in spring and autumn, generally about forty in the year' (*Annual Report of Liverpool Royal Institution*, 11 February 1876, p. 6 – Liverpool University Archives).

† A very poor report of this meeting appeared in the *Liverpool Daily Post* on 6 October (p. 6), and it is almost the only occasion on which that newspaper commented on university extension.

‡ *Liverpool Daily Post*, 7 October 1878. The Liverpool Record Office has registers of statistics for the free lectures from November 1865, but the sub-committee minute books do not begin until May 1879.

Lectures Sub-Committee thought this 'too special in its character' and asked for a continuation course on English Literature after 1720 for which Cambridge had no lecturer available. Instead Hall Caine, not yet a successful novelist, offered to lecture at a lower fee. No further courses were taken from the Syndicate by the corporation, although in 1881 there was a direct approach to a Cambridge lecturer at James Stuart's suggestion.*

In the same year that the corporation approached the Syndicate, a town meeting established a committee to work for the establishment of a University College in Liverpool. Like the Yorkshire College it began in 1881 with a scientific bias and with most of its students attending evening classes.† Nevertheless the Syndicate withdrew from Liverpool at this point to allow the University College to develop its own courses,[10] while continuing to work in Lancashire and Cheshire. In 1884 Liverpool became part of the Victoria University but continued to provide some extension courses independently. In February 1899 as a result of a lecture to the Literary and Philosophical Society a Society for University Extension (modelled on the London Society) was established.[11] Despite this early withdrawal from Liverpool some traces of Cambridge influence were still visible there some seventy years later.

The links between the Syndicate and Nottingham University College were much closer and lasted longer than elsewhere. The Nottingham College was established almost entirely as a result of the extension courses which began there in October 1873. The Mechanics' Institution had sponsored a University Extension Committee in addition to providing free accommodation.[12] The Committee required a large subsidy because the sale of tickets only met about one third of the total cost:

| | | | |
|---|---|---|---|
| Michaelmas 1876 | £103 | 4s. | 6d. |
| Lent 1877 | 94 | 16 | 0 |
| Michaelmas 1878 | 97 | 1 | 6 |
| Lent 1879 | 81 | 18 | 6 |
| Michaelmas 1879 | 94 | 11 | 6 |
| Lent 1880 | 58 | 16 | 0 |

* Liverpool Record Office, Min/Lib 3/1, 2–18. I am indebted to Prof. Kelly for telling me about this episode and its sources.

† *Liverpool University Recorder*, October 1970, pp. 10 and 11. Unfortunately Prof. Ormerod's typescript history (Liverpool University Archives) is incomplete for this period.

Michaelmas 1880     £127   9   0
Lent 1881             86   1   0*

In the absence of proper account books† it is difficult to discover how the subsidy was obtained. Some came from local industrialists and some from the Town Council. Richard Enfield, whose clerk Charles Abbott handled the financial arrangements, was the son and brother of former town clerks and it was probably through him that help was obtained from the corporation.‡

According to a local tradition Enfield and the organising lecturer, Moore Ede, discussed the need for an extension college in Nottingham late in 1874.§ Enfield soon found an anonymous donor who would give £10,000 for endowment if the corporation would erect suitable buildings. On 4 January 1875 Enfield wrote to the town clerk: 'I believe the buildings should at first comprise a lecture theatre seating about 500 or 600, two class-rooms seating about 150 or 200, a small room for a library, a chemical laboratory, and a residence for one resident lecturer; but that it should be planned with a view to subsequent extension.'[13] Enfield's principal supporters seem to have been the Rev. J. B. Paton and William Henry Haymann, a local manufacturer who was probably the anonymous donor.[14]

The corporation appointed its own university extension committee to negotiate with Enfield about the donor's wishes. A suggestion that the college should be erected near the Castle (which was about to become the Town Museum) was rejected in March 1875, but all other negotiations went well and in June a draft trust deed was agreed. In October the problem of a suitable site was referred to the public buildings committee, which was at this time laying out the Horse Fair Close in building lots.[15] It suggested a joint building on this site for the public library (which was urgently needed), a museum, schools of science and the extension college. This was accepted by the donor, the Syndicate agreed

* Nottingham Archives Dept., Acc. M 1726. The receipts include small sums for special classes and the sale of syllabuses. The three courses held each term must have cost between £200 and £300.
† With the exception of the volume cited above they are believed to have been destroyed in World War 2.
‡ In 1898 Enfield said that the Drapers' Company had assisted and the Town Council had given its profits from the Gasworks (*Report of a Conference*, 1898, p. 34 – BEMS 28/3).
§ A. C. Wood, *History of University College Nottingham* (Oxford, 1953), p. 15. Ede gave his first lectures in Nottingham in October 1874.

to the specification, and an architect was found by public competition.[16] The foundation stone was laid on 25 September 1877 by the mayor in the presence of Mr Gladstone.* When opened in 1881 it had cost about £43,000. Although University College moved to the outskirts of Nottingham many years ago, the elaborate Victorian Gothic structure still stands at the corner of Shakespeare and South Sherwood streets.

James Stuart was greatly concerned with the establishment of the college and in March 1876 he wrote a long letter to Richard Enfield about the 'ultimate consolidation of the scheme for University Extension'. This letter obviously continued a discussion which they had had previously and he referred to the difficulties caused by large numbers attending the classes and submitting written work. He rejected the view that minor economies would solve the problem and suggested instead a number of local colleges serving a group of small towns as well as the large town in which they were situated. Each would have a resident staff of lecturers who would move to other colleges from time to time. In this way the strain of travelling could be reduced for local lecturers.[17] Stuart became one of the Cambridge representatives on the committee of management in due course, and three of the four professors appointed were Cambridge local lecturers.† The Principal, and professor of language and literature, was the Rev. J. E. Symes, a friend of Bernard Shaw, a member of the Land Reform Union and a Christian Socialist, which caused some local difficulties.[18] Extension courses, which had continued without a break, were moved into the new building. The Town Council now took full responsibility and advertisements were issued by the Town Clerk under the heading 'University Education for the People'.‡ In 1883 Nottingham received the privilege of affiliation with Cambridge which it had first sought in 1878.§ University College organised courses locally for the Syndicate and some of its

* The University College sub-committee met almost daily before the ceremony and had the platform enlarged as late as 21 September (Nottingham Archives Dept., misc. committee minute book, 1875–81, pp. 209–36).

† The fourth came from London. Wood, *University College, Nottingham*, pp. 25 and 33.

‡ *Nottingham Journal*, 4 October 1878. The later accounts for tickets sold were audited by the Borough Accountant.

§ University Archives, C.U.R. 57.1, 83. Tenth Annual Report (BEMS 22/1, f. 78). Prof. Kelly (*Adult Education*, p. 230) was misled by the length of time taken to obtain the affiliation statute and thought that Nottingham was never affiliated to Cambridge.

own courses were approved by the Syndicate, but it soon became obvious that the more serious students preferred a London external degree to a Cambridge Certificate.[19] By 1882 equal prominence was being given in the college *Calendar* to London and Cambridge courses, and from 1895 the Vice-Chancellor's Certificate regulations were not included.[20] Cambridge affiliation courses were held at Nottingham until 1907 and terminal courses until 1914, but the numbers attending were comparatively small and the chief aim of the students was a London external degree or a teacher's diploma.

A less permanent venture with which the Syndicate was associated for seventeen years was the Crystal Palace Company's School of Art, Science and Literature. After Paxton's exhibition buildings had been removed from Hyde Park to Sydenham the Company started a number of projects to encourage their use, including pleasure gardens and an 'educational institution'.* The latter opened about 1860 and by 1877, when the first approach was made to the Syndicate, it had ladies' and gentlemen's divisions and a school of practical engineering.[21] James Stuart inspected it for the Syndicate and in the same year the first courses were given.† Thomas Hughes, the author, was chairman of the Company's Educational Committee at the time and on 3 June 1878 he signed a memorial to the University seeking the privileges of an affiliated college.‡ Although the memorial expressed the hope that it would become 'a Collegiate Institution for South London', the Syndicate did not consider its status sufficient for affiliation in 1883 when the new statute had been approved.[22] Nevertheless the Sydenham centre (as it was usually known) continued to take Cambridge courses and examinations until 1893.§

During this period and until 1891 the Secretary of the Syndicate

---

* The London, Brighton and South Coast Railway opened a station there in 1854 and the London, Chatham and Dover Railway its rival High Level station in 1865 (E. Course, *London Railways*, London, 1962, p. 233).

† *Reminiscences*, p. 207. In Michaelmas term 1877 fifty-five students attended courses in chemistry and political economy.

‡ University Archives, C.U.R. 57.1, 85. Hughes was lecturing at the Working Men's College at this time, but his work for the Crystal Palace Co. does not seem to have been recorded. See E. C. Mack and W. H. G. Armytage, *Thomas Hughes* (London, 1952), chapter 10.

§ The last Cambridge course there was given in Lent 1893 (*20th Annual Report*, 1893). It is not clear from the Board's records why the Company then ceased to take the courses, and the Company's records were apparently destroyed by fire.

was George Forrest Browne. He was born at York in 1833 (ten years before Stuart). His grandfather had been sub-chanter in the cathedral there and through him Browne's father had become a proctor in the archbishop's courts. Until the decline of the ecclesiastical jurisdiction, Browne himself was intended to be an advocate in the courts.* However his father realised that there was no future in this career and decided to send his son to the university and to invest his savings in George Hudson's railway schemes.† When the latter collapsed, it became necessary for G. F. Browne to go to St Catharine's College which had a number of Yorkshire scholarships. An unfortunate accident prevented him from getting a good degree, so he left Cambridge for the newly-founded Trinity College at Glenalmond to teach mathematics and classics. He afterwards claimed that this removal to the Scottish episcopalian college was the turning point of his life because it made him work.[23] While still in Scotland he was ordained by the Bishop of Oxford,‡ and in 1863 he returned to Cambridge as chaplain and fellow of St Catharine's. For the next thirty years he immersed himself in university politics. His original interest was scientific and in 1865 he wrote *Ice Caves of France and Switzerland*, but he later became interested in Anglo-Saxon history and was elected Disney Professor of Archaeology in 1887. However his main work for the University was as an administrator. He was a member of the Council of the Senate for most of the period between 1874 and 1892. He was a proctor from 1867 to 1869 and from 1877 to 1881. He was secretary of the Royal Commission on the University which was appointed in 1877. He founded the *University Reporter* in 1870 and edited it for twenty-one years.[24] In 1869 he was appointed Secretary of the Local Examinations Syndicate to which he added the Local Lectures Syndicate in 1876. In addition he helped to reorganise the local Conservative party. His autobiography with its coy references to himself in the third person§

* A proctor was the equivalent of a solicitor and an advocate of a barrister in ecclesiastical courts. By the mid-nineteenth century it was clear that the most profitable parts of the ecclesiastical courts' jurisdiction would be abolished.

† G. F. Browne, *The Recollections of a Bishop* (London, 1915), pp. 1–21. Much of the description of Browne's career is taken from this source.

‡ By letters dimissory because the Anglican church is not established in Scotland.

§ He was influenced to do this by the story, quoted in his preface, of the Cambridge don who exhausted the University Press's entire stock of upper case I.

gives the impression that he was the Pooh-Bah of the University. Unfortunately his memory cannot be trusted in some of his recollections. In 1891 he left Cambridge to become a canon of St Paul's and in 1897 he became Bishop of Bristol. In 1914 he retired to London where he died sixteen years later.

At first Browne held his two secretaryships separately, but it became clear that it would help Local Lectures financially if the two Syndicates were combined. On 22 February 1878 the Local Examinations Syndicate appointed a sub-syndicate of three to 'consider and report on the arrangement of the work and the remuneration of the Secretary'. The report, presented on 10 May, recommended amalgamation of the two Syndicates with a secretary at £700 a year and an assistant secretary at £200: 'The duties of the Assistant Secretary would be to assist generally the Secretary. During the first half of the Michaelmas Term and during the Easter Term, his services would be required about two hours daily; but during the latter half of the Michaelmas Term, and the greater part of the Lent Term, his whole time would be required.'*

The Local Lectures Syndicate does not seem to have been consulted, but no opportunity was lost to emphasise its difficult financial position,† and ten days later it agreed 'that steps should be taken to unite the Local Examinations and Local Lectures Syndicates, and that the Vice-Chancellor be requested to communicate this Resolution to the Council of the Senate'.[25]

An assistant secretary was appointed on 3 June, but the joint Syndicate did not take over until the beginning of Michaelmas term. One of its first actions was to appoint a sub-syndicate (including James Stuart) 'to draw up a scheme for the management of the combined Syndicate',[26] but no such scheme has survived.

In practice the place of the Local Lectures Syndicate was taken by the Local Lectures Committee which was allowed a great deal of independence and reported only briefly to the Syndicate. It also submitted a separate annual report to the Senate with the slightly odd result that there is a series of *Annual Reports of the Local Examinations and Lectures Syndicate* printed at Easter which are concerned with examinations only, and a separately numbered

---

* Local Examinations Syndicate, minute book 2, 10 May 1878. There was also to be a second clerk – 'a good shorthand writer'.

† Local Examinations report stated that the 'Local Lectures Fund may be taken as just sufficient to meet expenses'.

series of *Annual Reports on the Local Lectures of the University* which were printed in June. Unfortunately there are no committee minutes for the whole of Browne's secretaryship – a retrograde step for which he must be held responsible.* Browne seems to have been indifferent to the cause of Local Lectures. His official policy was stated by the Syndicate in 1879 to be 'the consolidation of the work at a few main centres',[27] but in a less guarded moment he defended it because it was a source of employment for unemployed Cambridge graduates.† Perhaps his chief contribution to the work of Local Lectures was the provision of a permanent office for the Syndicate. When Browne took over the lectures work, the original office was apparently given up and Browne's rooms in St Catharine's were used for this as well as Local Examinations work. By 1884 he had rented a second set of rooms there and a college lecture room was being used to pack examination papers. Between six and eight people were then employed there at busy periods.[28] Because the rooms had to be given up, the Syndicate obtained permission in March 1884 to buy Finch's Buildings in Guildhall Place out of its accumulated examination profits. There was some criticism of the inaccessibility of this site, so the Syndicate bought 2, 3 and 4 Mill Lane and built new offices there.‡

The continued existence of Local Lectures depended on the work of three very active assistant secretaries to whom Browne delegated most of the work –

> The Rev. William Cunningham, 1878–1881
> Dr R. D. Roberts, 1881–1890
> Stanley M. Leathes, 1890–1892§

William Cunningham was another Scot who, like James Stuart, had migrated from a Scottish university to Cambridge. His

* BEMS 5/1 begins the series of minute books in February 1892 with very rough notes. The minute book lettered 'Non-Gremial Syndicate' and mentioned by Stuart in his *Reminiscences* (p. 170) as having been bought by him is in fact the second Local Examinations minute book (1866–86) which he did not buy.

† *University Reporter*, 1891, pp. 195, 196. I have found no evidence for Miss Lemoine's assumption that Browne resented the independent existence of the Lectures Syndicate ('The North of England Council', Manchester M.Ed. thesis, p. 290a).

‡ University Archives, C.U.R. 57.1, 121. Roach, *Public Examinations in England 1850–1900*, pp. 166, 169. 'Syndicate Buildings', as it was known, still stands next to Stuart House, but is used for other university purposes.

§ The *Historical List of Secretaries and Lecturers* published by the Syndicate in 1923 gives different dates for all three, but the Syndicate minute books are quite clear about the appointments and resignations.

family were Free Presbyterians who had left the Established Church at the Disruption. Cunningham, however, was attracted to Anglicanism and was permitted by his parents to go to Cambridge for this reason. In 1872 a scholarship enabled him to move from Caius to Trinity where he became a friend of Stuart and Stanton. Stanton was a particularly close friend and Cunningham contributed to his *Tatler in Cambridge* as 'Parson Bill'.[29] He was ordained at Ely in 1874 and, having failed to get the expected fellowship at Trinity, he became one of Stuart's local lecturers. In January 1874 he took the place of Moore Ede at Bradford and Leeds as lecturer in Political Economy and in October he went to Liverpool to lecture on that subject and on Logic. He was very successful as organising lecturer there, but his health was poor and he welcomed the offer to return to Cambridge as Browne's assistant secretary in June 1878.[30] In February 1880 he offered his resignation because of ill health, but he was asked to continue with a promise of temporary assistance. A year later he again submitted his resignation in order to go to India to recuperate and this time it was accepted.[31] He continued to be interested in university extension, but his subsequent career in the University and the diocese of Ely prevented him from taking a very active part for the rest of his life. His best known work, *The Growth of English Industry and Commerce* (1882), owes something to his lectures for the Syndicate.

The next assistant secretary, Robert Davies Roberts, was an enthusiast like James Stuart. He had no doubts about the value and importance of university extension and worked at Cambridge (where he was assistant secretary from 1881 to 1890 and secretary from 1894 to 1902) and at London (secretary from 1885 to 1894 and from 1902 until 1911) to make it a success. Roberts was the son of a Welsh timber merchant and was born at Aberystwyth in 1851. His family were Welsh Calvinistic Methodists and his mother was a descendant of Thomas Charles of Bala who had revived the Welsh Circulating Schools.* He went to school at Aberystwyth, Oswestry and the Liverpool Institution. In 1867 he went to University College London where he obtained a first

---

* The Calvinistic Methodist Church of Wales is also known, in English, as the Presbyterian Church of Wales. Roberts' uncle was Principal of the Church's theological college at Trevecka from 1888 to 1891 (W. P. Jones, *Coleg Trefeca 1842–1942*, n.d., pp. 92, 93).

class degree in geology. Like Stuart he then moved to Cambridge to obtain a degree in the Natural Science Tripos. In this he was interrupted for a time by his father's death, but he obtained a first in 1875.[32]

Although a member of Clare College, Roberts was soon enlisted as a local lecturer by Stuart. In Michaelmas 1875 he opened a centre at York with a course on physical geography. Because there was difficulty in obtaining a second lecturer, Roberts repeated his course in the evenings.* He was so popular at York that he was asked to give a course on geology the next year, which he did despite poor health.[33] In November 1875 he even wrote to Stuart asking permission to repeat his lectures to local schoolchildren *gratis*. Stuart reluctantly refused because it might lead to difficulties with the local committee.[34]

For the next two years Roberts did no lecturing for the Syndicate. He had returned to Aberystwyth as a temporary lecturer at the new University College. But while there he gave the first course of extension lectures under the auspices of the College. He began on 24 October 1876 with a statement of his beliefs: 'In my most sanguine moments I find myself looking forward to the time when it will be considered as necessary to have in every town and district educated teachers of the people as it is now to have pastors to look after their religious education. One of the functions of the University College of Wales is to stand forth as a witness ever before the people that they must not rest until the means of higher education is within the reach of all.'†

When his lectureship came to an end in 1878 he returned to Cambridge with his newly-acquired London doctorate.‡ He lectured for the Syndicate at the Crystal Palace in 1878–9 and at Barrow in Furness, Carlisle and Lancaster in 1879–80. On 12 March 1881 the Syndicate appointed him 'their Assistant Secretary for the special purposes of the Local Lectures' at an annual salary of £50, on the proposal of Professor Stuart.[35] Although he

* *Calendar of Local Lectures*, 1880, p. 139. This seems to have been the first time that the same course was given during the day and in the evening.

† Quoted by Thomas, *Harlech Studies*, p. 2. Thomas' claim that these were the first extension lectures in Wales is untrue. Cambridge had established centres in South Wales in the previous year.

‡ The existence of which Cambridge ignored. This was the first degree of doctor of science to be gained by a Welshman. Roberts' departure from Aberystwyth was probably caused by the College's financial difficulties. See R. B. Knox, *Wales and 'Y Goleuad'* (Caernarvon, 1969) pp. 158, 159.

replaced Cunningham the appointment was on a different basis, and for the first time since Browne became Secretary contact was established with the local centres. The Syndicate 'are enabled to hold direct personal communication with the towns in which lectures are held. They will be able also to provide short preliminary courses of lectures and classes, with a view to explain their method, in towns which desire to bring themselves into educational relations with them'.[36] This reversion to Stuart's methods was probably the result of Oxford's moves to establish university extension courses.

Roberts took up his new duties with great zest. The greater part of the next *Annual Report* is taken up with his report to the Syndicate on his travels:

Between July, 1881, and Easter, 1882, I visited all the centres at which Lectures have been given this winter . . . also Liverpool when negotiations for a course of lectures next term were going on. I further went by invitation to other places where the Lectures had been discontinued, but where there was a move towards again taking the matter up . . . I visited many of the first-named seventeen centres, which are under this Syndicate, more than once at the request of the Local Secretaries, either to be present at a Committee Meeting, or at the first Lecture of the Course. From what I have observed I believe the occasional presence at a Committee Meeting or a Lecture of some one representing the Syndicate is a source of encouragement to the Local Committees.[37]

In the following year Roberts' report to the Syndicate had grown so large that it was published separately two months before the *Annual Report*.[38] That year he had made 46 visits to centres and in the following year he made 62. He soon summed up the situation from his visits and in his second report clearly summarises the difficulties of the work and obstacles to its progress:

1. *Local financial difficulties.* Nothing has been brought out more clearly by the experience of the past ten years than the fact that the great obstacle to the wider extension of the Local Lectures Movement is not the absence of a demand for or of interest in education, but the difficulty of obtaining funds to meet the expenses . . . This difficulty presses most heavily upon just that section of the community which would be most benefited by the adoption of the scheme, viz. the wage-earning population.

2. *Need for some University recognition of the Vice-Chancellor's Certificate* . . . I have frequently been asked, especially by working men, what value the Vice-Chancellor's Certificate possesses, and they have expressed

disappointment on learning that the Certificate has no University recognition, and confers no University privilege . . .

3. *Need for some training for Lecturers.* This is a matter which will assume more and more importance and become more urgent as the work of the Syndicate grows and the demand for new Lecturers becomes larger. Special qualifications are needed to make a successful University Extension Lecturer . . . [39]

Roberts continued his work of travelling secretary and in 1883 his appointment was confirmed by the University at a salary of £200 a year. The following year he also became university lecturer in geology and fellow of Clare. He resigned his lectureship after a year because the London Society for the Extension of University Teaching asked him to become their secretary. The Society was a voluntary body with representatives from Oxford, Cambridge and London on its committee. Because its work was largely confined to the London postal district there was no difficulty about combining his new post with the assistant secretaryship at Cambridge.\* A sub-syndicate was appointed which reported favourably on making the experiment. Roberts proposed to spend two or three days a week in London in term time and to spend the vacations visiting Cambridge centres. The sub-syndicate felt that some adjustment in salary might be required,[40] but none was made before his resignation in 1890.†

Roberts' relationship with G. F. Browne was probably uneasy, although this was never obvious in public. Robert's biographer wrote that '[Browne] was a conservative and cautious in temperament. [Roberts] was liberal and fervid. Fortunately, they were bound by mutual respect and a spirit of toleration. The assistant admired his chief's skill in steering his modified schemes through university committees'.[41] But perhaps the chief reason was that Roberts was Lecture Secretary in all but name. Two bundles of correspondence which survive for 1885 and 1886 illustrate this point very well. The Syndicate's notepaper enjoined all correspondents to address their letters to Browne, but all the lecturers and most of the local secretaries wrote direct to Roberts. Those

\* Local Examinations Syndicate, minute book 2, 8 March 1883 and 24 October 1885. The minutes make it clear that the invitation from London came in 1885 and not 1886, as is usually said.

† He was succeeded by Stanley Mordaunt Leathes, an historian from Trinity, who served for two years before becoming assistant registrary. In 1903 he became secretary of the Civil Service Commission.

who knew the administrative arrangements wrote to Browne only on the most formal occasions or when they enclosed fees. Browne was probably too busy to spare the 'two or three hours a day' which he asked Cunningham to devote to these duties,[42] but he was not particularly interested in extension work. In his autobiography this aspect of his work occupies only three pages and these contain several mistakes.* He was a member of the Local Lectures Syndicate from its beginning, but his name rarely appears in the first few years. He did attend the public meeting at Liverpool in January 1874, but his contribution was negligible.[43]

But perhaps the most conclusive evidence of Browne's indifference is the figures published in the *Annual Reports*. In 1875 and 1876 there had been thirty Cambridge centres and fifty courses. When Browne took over in 1877 the number of centres fell sharply to twenty and the number of courses to thirty-six. In the following year the numbers were seventeen and thirty-two. There they remained until R. D. Roberts' appointment began a slow revival. It was not until 1882 that Cambridge again had thirty centres and the total number of courses remained at thirty-seven.† Through Roberts' enthusiasm the numbers continued to expand, but the initial enthusiasm under James Stuart had been lost when Browne reduced the numbers at the beginning of his secretaryship. Most centres now took only one course a year or a term, which made it almost impossible to provide series of courses leading to a Vice-Chancellor's Certificate.‡

From the beginning of his secretaryship Browne seems to have emphasised the financial aspect of the work at the expense of the educational. By insisting that each centre must pay for courses in advance and because he provided no encouragement for centres in temporary difficulties, he rapidly reduced the number of centres. In his first year centres defaulted to the extent of £112 2s 4d and the auditors found a deficit of £9 10s 9d which Browne agreed to deduct from his stipend.[44] For the next few years no accounts have survived and when they begin again in 1879 they are more difficult to understand than those kept by Stuart. For the first two years the account book contains little more than memoranda:

* *Recollections*, pp. 128–30. Not the least of his mistakes is the claim that he was co-founder of university extension.
† See Table 1.
‡ Nottingham continued to take several courses each term, showing what could be done if there was sufficient encouragement.

Mr Read has not been paid* for the July papers at Nottingham.
The other have (June 21, 83).
* the 2.10.0 & 10/-.
paid Aug. 25th 83*

From Michaelmas 1881 to Lent 1883 there are also separate
pages devoted to each centre and each lecturer, but there are no
balance sheets.† However it seems that Browne had eliminated the
annual loss, but at the cost of reducing the amount of work done
where it was most needed.

James Stuart had assured the University that it would not be
required to find any money for Local Lectures in order to get his
scheme accepted. He was also faced with the argument that
Local Examinations were able to make a profit. At first he may have
hoped that the money could be provided locally. 'Before the
Syndicate can enter into business arrangements with any particular
locality, two conditions must be fulfilled; first, a competent local
committee must be appointed with a Secretary to correspond with
the Syndicate; and, secondly, a guarantee fund, sufficient to meet
all expenses, must be secured from local sources.'[45]

He was probably influenced in this by his experiences with the
North of England Council. Its problem had not been finance, but
a shortage of suitable lecturers. The 'young men engaged in
commercial and professional pursuits' and even more the 'working men' for whom the Syndicate made provision were not in the
same fortunate financial position. In any town the full fee was
more than they could afford and in the smaller towns it was
difficult to raise a sufficiently large subsidy. To some extent Stuart
recognised this while he was still Secretary. He persuaded Trinity
College to alter its statutes to allow fellows to work for the
Syndicate.‡ After Stuart ceased to be Secretary he obviously
became convinced that a permanent subsidy in some form was
required, but it was then too late to take the necessary steps.

The first year's accounts of the first three Cambridge centres
show how difficult it was to be self-supporting. Derby in its first
year had a deficit of £264 14s 2d.[46] At Nottingham the total cost

* BEMS 18/3, 18. The entry appears to relate to payment for a Local
Lectures examination in 1883.
† Although the account book continued in use until 1888 the later accounts
are more confused.
‡ G. F. Browne as Secretary to the Royal Commission was in a good position
to further the same alteration to the statutes of other colleges, but did not.

of the lectures was £534 19s 10d and the total receipts £295 3s 9d. Donations of £346 converted a serious loss into a balance in hand of £106 3s 11d.[47] Yet Nottingham had the advantage of free rooms for most of its courses. The cost of lecture rooms was the cause of failure of the first Leicester centre. The hall at the Working Men's College was free, but was not used after the first term because the committee refused to admit women.[48] The Museums Committee allowed the use of their hall, but minuted that 'the terms stand over and be decided upon at the next meeting of the Committee'.[49] Unfortunately nothing more was done for ten months and the chairman apparently gave reason to believe that there would be no charge – the curator was secretary of the Local Lectures Committee. On 30 April 1874 the committee imposed a charge of half a crown a night and in August five guineas was paid off the arrears. The lectures' accounts produced the following month apparently included the total sum due to the Museum Committee:

| Receipts | | | | Payments | | | |
|---|---|---|---|---|---|---|---|
| Subscription tickets | £125 | 6 | 0 | Lecturers | £220 | 11 | 9 |
| Guarantors | 130 | 5 | 0 | Travelling expenses | 25 | 19 | 3 |
| Students' fees | 47 | 13 | 2 | Local expenses | 67 | 10 | 2 |
| | 303 | 4 | 2 | | 314 | 1 | 2* |
| Owed to bank | 10 | 17 | 0 | | | | |
| | 314 | 1 | 2 | | | | |

In November and December the Museum Committee instructed the Town Clerk to demand the balance due. In June 1875 a further bill, including use of the hall in the second session, was sent, but there is no record of a further payment until October 1875 when another five guineas was received. Apparently no further payments were made, and the centre closed in 1876.[50] Similar stories could probably be traced elsewhere. At Halifax in 1877 'the expenses of the experiment exceeded the sum guaranteed'; at Keighley in 1875 'the lectures here were well attended, but the Committee found that their funds did not allow them to enter upon a third year'; and at Lincoln in 1876 'the Committee being unwilling to call heavily upon the Guarantors decided to discontinue the lectures'.[51]

Under Browne each local centre had to struggle on in the best way it could. Much depended on an enthusiastic local committee

* Reconstructed from figures given in *Leicester Journal*, 25 September 1874.

and, in particular, on a good local secretary. Both James Stuart and R. D. Roberts saw the need to encourage local secretaries by frequent meetings, but Browne in 1887 was able to say that he knew most of the secretaries only by their handwriting.[52] Both Stuart and Roberts believed in calling conferences as well as visiting local centres, but there were no meetings called by the Syndicate between 1875 and 1887. Local centres were left to struggle on with very little guidance. Since one unsuccessful course could wipe out any surplus and close down the centre, they were usually very cautious. Well-tried lecturers with popular subjects were booked if possible and there was rarely a balanced programme such as Stuart had tried to provide.

It is difficult to find any other element in the success or failure of centres at this period. As we have seen Derby failed after two years of lectures through lack of an enthusiastic committee. Chesterfield, in the same county and with little more than a third of Derby's population, was successful from its beginning in 1875.[53] Southport with only a fraction of the population of Bolton and no industries was far more successful, although both centres began in the same year. Even the existence or non-existence of other adult education bodies in the town does not seem to have affected the success of the Cambridge courses. Southport had a Ladies' Educational Association and a Scientific and Literary Society, but Chesterfield does not seem to have had even an effective mechanics' institute at this time. Even the availability of a suitable lecture hall could not save a centre. Burslem had the Wedgwood Memorial Institution (so lovingly described by Arnold Bennett[54]) but the lectures there were a complete failure. Only at centres where the local committee had its own building, as at Nottingham and Chesterfield, did this factor influence their success.

When the Leicester centre was revived in October 1882 all the enthusiasm seems to have come from the town and not the Syndicate. The Rev. James Went had moved from Huddersfield, where he had been headmaster of the grammar school and secretary of the local lectures committee, to Leicester to be headmaster of the recently-founded Wyggeston Boys' School. His enthusiasm persuaded the Leicester Literary and Philosophical Society to subsidise a course in English history from the Revolution to the Reform Act.[55] A strong local committee was formed with local Anglican and nonconformist clergy well represented. No one

84

from the Syndicate visited Leicester to discuss the arrangements and it is probable that Browne's correspondence was restricted to the amount of fees to be paid.* Despite the approval of the Working Men's Club, the first lecture had only 'a moderate audience' and for the second the room was but 'Fairly filled'.[56] The lectures were given in the afternoon and evening to save costs, but the Society made a loss of £23 2s and decided not to hold a second course in the spring.[57]

By this time the Syndicate were 'no longer alone in the field of work entrusted to them'.[58] In 1876 a public meeting in London had led to the foundation of the London Society for the Extension of University Teaching. This was not part of the University of London, but included in its council representatives of the London colleges and of the three universities.† It continued work until 1902 when the reorganised University of London established a Board to take over the work.[59] In 1878 the University of Oxford began to provide extension courses in competition with Cambridge, but it was not until 1885 that this seriously affected Cambridge or until 1892 that the Delegacy for University Extension was created.[60] Oxford offered short courses of six lectures followed by an examination, to the great annoyance of the Syndicate which felt its work was being undermined.[61] In 1879 the University of Durham began extension work in the North East, but three years later were obliged to enlist the Syndicate's support.[62] For the next fourteen years these courses were regarded as joint, although the Syndicate seems to have done most of the work. In 1895 Durham asked to withdraw from the agreement and run independent courses once more, but Cambridge retained most of the centres in Northumberland and Durham.[63] The federal Victoria University also provided some extension courses from 1886.[64] All these rivals helped to keep the number of Cambridge courses below Stuart's maximum until 1891.

* Copies of his letters to local centres for the previous year have survived (BEMS 12/1). They are confined to requests for payment.

† The London Society had a monopoly in the London postal district, but also organised courses in the Home Counties. For the dispute about the appointment of representatives from Cambridge see University Archives, C.U.R. 57.1, 70.

# 6

# A. Berry and R. D. Roberts, 1891-1902

The Syndicate recognised the unsatisfactory nature of the administration under G. F. Browne by the alterations which it introduced immediately after his resignation in 1891. It decided that in future there would be two secretaries of equal standing – one responsible for lectures and one responsible for examinations. On 14 March 1891 the Syndicate decided to offer the examinations post to Dr J. N. Keynes,* and the lectures post to Dr R. D. Roberts. Each was to have his own assistant secretary.[1] Unfortunately Roberts felt unable to leave the London Society at this point and declined. On 10 May the Syndicate chose Arthur Berry in his place. Berry was a younger man than Roberts, but had also been at University College London before moving to King's. He was a mathematician who had lectured on science for the Syndicate since 1886. Berry's part in the history of the Syndicate has never been properly appreciated. He combined a strong enthusiasm for university extension with an active participation in University affairs.† By this means he was able (probably in consultation with Roberts) to introduce many urgently needed reforms with a minimum of fuss. He introduced the first committee minute book and obtained a resolution that minutes should be kept in future.[2] He introduced letter books and a separate series of accounts, and he took steps to improve the status of lecturers.[3] He almost certainly introduced proper procedures for the appointment of lecturers, circulated lists of available lecturers and provided printed registers for the courses.‡ He continued Roberts' practice of visiting local

---

[1] For References see p. 201.

* The father of Maynard and Geoffrey Keynes.

† See Oscar Browning's correspondence with him in the Browning papers at Hastings Public Library for this. Berry and Browning were friends until Browning quarrelled with him about college politics. See also *Annual Report of King's College Council*, 1929, p. 2.

‡ A bundle of printed forms which are labelled as having been issued during

centres frequently and the Syndicate issued his reports on this separately. His secretaryship began very successfully, yet despite all his enthusiasm and efficiency it ended with a reduction of about half in the work done by the Syndicate. The reasons for this were, however, completely beyond his control. In 1891 and 1892 there was a tremendous demand for scientific and technical courses from the new county and county borough councils which fell away completely before Berry's resignation in 1894. On the other hand Berry saw the beginnings of what eventually became the University of Exeter and persuaded the Syndicate to try to get a government subsidy for all extension courses.

Berry was obviously a close friend of Roberts and it is possible that he became secretary to ensure that the work was carried on along Roberts' lines until the latter could return to Cambridge. When he resigned in October 1894 no reason was given and the Syndicate quickly decided to offer the post to R. D. Roberts again.* Roberts held out for a better salary and then accepted. Berry spent the rest of his life at Cambridge, becoming university lecturer in mathematics and Vice-Provost of King's. He was a member of the Syndicate for part of the time and continued to give extension courses. When he died in August 1929 his slides were bequeathed to the Syndicate.[4]

Although the need for technical education in Great Britain was debated from the first half of the nineteenth century, little was done beyond the South Kensington classes until 1881. Then A. J. Mundella appointed a Royal Commission which made definite recommendations for a national system. Further progress was delayed by the need for local government reform. County councils and county borough councils were created in 1889 and one of the first powers they received was the right to levy a penny rate for technical education.[5] Since elementary education was not the concern of either authority and the product of a penny rate was not usually large, it is not surprising that little was done at first. However in the following year the Local Taxation (Customs and Excise) Act provided comparatively large sums of money from

his secretaryship survives (BEMS 38/9). Unfortunately there are no minutes about their introduction and no completed registers have survived.

\* Local Examinations Syndicate, minute book 4, pp. 46 and 52. There is a College tradition that at this time Berry undertook most of the college teaching in mathematics to allow his colleague, H. W. Richmond, more time for original research. (Information provided by Mr John Saltmarsh.)

the excise duty on spirits which could be used either to relieve the rates or to subsidise technical education.\* This grant – inevitably to be known as 'whiskey money' – was used by most of the councils for the latter purpose.† Many turned to the Cambridge Syndicate and the Oxford Delegacy to provide the necessary courses and both were besieged with requests for scientific lectures.‡

The grant was not received until 1891 and on 15 July 1891 Kent County Council appointed a Technical Education Committee to spend about £22,500 which it hoped to receive. When the Committee met on the same day it already had a request from the South East Counties Association for University Extension for a grant.§ A week later the Association forwarded a printed memorandum describing the work already being done in Kent and seeking a subsidy for its expansion:

University Lecturers are prepared to travel from town to town and village to village wherever Centres can be organised, in School Rooms and Institutes, and give Courses on the following . . . Chemistry applied to Common Life; Soils, Plants and Animals; Agricultural Chemistry; Science of Agriculture; Entomology; Botany; Light and Heat; Elementary Geology; Astronomy; Physical Geography; Laws of Health; Animal Physiology; Mechanics; Domestic Architecture; Industrial History; Political Economy; Electricity in the Service of Man, etc., etc.

These Lectures would be extensively illustrated and popularized by the use of coloured diagrams and lantern slides.[6]

The Committee allocated £3000 a year for extension lecturers, the Association appointed two organising agents for Kent, and sixty centres were quickly established.[7] The work was divided evenly between Oxford and Cambridge. In Michaelmas term 1891 the Syndicate provided 21 full courses and 19 short courses, but this dropped to only ten full courses a year later.[8]

---

\* 53 and 54 Vict., c. 60. The Government originally intended that this money should be used to buy up redundant public houses. An outcry from the temperance movement prevented this and as a compromise it was offered to the new councils (B. Webb, *Our Partnership*, London, 1948, pp. 76, 77).

† The county councils usually shared it with local authorities within their area.

‡ At the end of 1890 the Syndicate agreed to notify local centres of the possibility of obtaining grants (Local Examinations Syndicate, minute book 3, 29 November 1890). The first edition of Mackinder and Sadler's *University Extension* (1890) emphasised the help which university extension could give to local authorities.

§ Kent Record Office, C/MC 12/1/1, 1–11. The Association was a union of Oxford and Cambridge centres in Kent, Surrey, Hampshire and Sussex.

Since the Committee paid all the Cambridge fees, only the cost of hiring a hall and advertising the course had to be found locally and all except five centres were successful. But by December 1891 the Committee was already complaining that local lecturers charged half the fees of Cambridge and Oxford lecturers, and an alternative proposal for 'instruction on Agriculture by Local Teachers' was put forward.[9] The secretary of the Association pointed out that the new scheme was suitable for 'big lads from 12 to 17 years of age', whereas the extension lectures were intended for 'miscellaneous audiences of grown up people comprising Gentry, Farmers, Shopkeepers, Artisans, Labourers, as well as the more intelligent young people'. As an alternative he suggested that they tried the 'Norfolk Experiment',* but instead the number of extension courses was reduced, and on 7 November 1892 the Committee decided to take the organisation of the courses into their own hands.[10]

The Syndicate also provided courses for Devon County Council. As early as October 1890 the county council had been approached by the National Association for the Promotion of Technical and Scientific Education. Mr Macan, a supporter of the Cambridge courses at Exeter, put forward a scheme by which the council would give a subsidy of between £3 and £10 to 36 towns to provide courses.

Resolved that the fees for the Lectures and Classes be left to the discretion of the organizing Local Committee in each case, so that they do not exceed the following amounts:

For those who have passed the 5th Standard   1s.
Others   2s. 6d.

That all fees be paid in advance and that the public be admitted to the Lectures on payment of a sum not exceeding 6d. each for every Lecture. Resolved that in each centre not less than 10 Students should be required before any course shall be commenced.[11]

This scheme was probably far less satisfactory than that for Kent because the smaller towns (which needed the greater subsidy) got the least, and those who had not reached standard 5 paid higher fees. Nevertheless the Technical Education Committee

* In Norfolk extension lecturers trained local teachers at urban centres, and they in turn conducted classes in rural areas (see below, p. 90). This was similar to the 'Guildford Experiment' which the South East Association had sponsored in 1889 (BEMS 22/1, 117).

adopted the scheme and Mr Macan's services. They were soon quarrelling with him about a circular which he had issued to local centres and had disclaimers printed in the Exeter and Plymouth newspapers.* Eventually this difficulty was smoothed over, but then the Committee discovered that local grammar school teachers could provide courses of twelve lectures for £30, whereas Cambridge and Oxford charged £45.[12] Cambridge provided eight courses on elementary chemistry and five on experimental mechanics in Lent term 1891. At Michaelmas five centres had elementary chemistry and five 'plant and animal life'. By Lent 1892 there were only seven Cambridge courses ('manures and foods' or 'soils, plants and animals') and after that none.[13]

The 'Norfolk Experiment' is unfortunately not so well documented.† Members of the Norfolk County Council knew of the Cambridge courses which had been given at Norwich since 1877. The chairman of the County Council afterwards described how, having persuaded his colleagues to embark on a policy of training teachers to give the lectures, he went 'in a state of great agitation to the Syndicate' to ask it to provide the training.[14] Under the scheme the county was divided into three districts and courses were given to schoolteachers at Norwich, King's Lynn and Elmham on Saturdays. The first course was 'Chemistry with special reference to Agriculture' and practical work was substituted for the classes. The County Council paid all the fees and expenses of those attending (including students' train fares). About 130 attended, of whom about 20 were expected to be able to conduct evening classes in the villages.[15] The course was repeated to smaller numbers in 1892, and a similar series was given for Lindsey County Council at Brigg, Lincoln and Louth to about 100 teachers.[16] Laboratory work at Cambridge was also arranged by the Syndicate for Norfolk teachers, and the work continued, although with smaller numbers, for many years more. A leaflet describing the work in Norfolk was issued by the Syndicate in 1894 for circulation to other local authorities,[17] and in April of the previous year it organised a two day conference at Cambridge for county councils. At this the Norfolk experiment

---

* Macan later became Secretary for Technical Education in Surrey where he quarrelled with the South Eastern Association for University Extension (*University Extension Journal*, vol. 1, pp. 123, 124).

† The County Council's technical education committee minutes do not begin until 1893.

and the co-operative scheme in Cambridgeshire were the prize exhibits.[18]

Cambridgeshire made a late start with technical education because of the illness and death of its first organising secretary.[19] After approaching all the adjoining authorities with a scheme for a joint organisation, the county council finally joined Cambridge Borough Council in organising all its work through the Technical Institute which the latter established in 1893.* The borough council had quickly decided that its share of the 'whiskey money' could best be spent on providing a building 'where nearly all the agencies concerned in the work of Technical Education can be concentrated and fresh agencies called into being'.† The entire grant of £1003 for 1891-2 was used to buy a building in East Road, and this was made available for extension courses too.

The County Committee asked the Syndicate to provide courses 'which have some direct relation to the lives and occupations of the people, such as Veterinary Science, Dairy Work, Fruit Culture, Vegetable Culture, Bee Keeping . . .'. It paid all the lecture fees, but each local committee was responsible for the transport of the lecturer 'from the nearest convenient station' and for local expenses.[20] Seventeen centres were organised outside Cambridge and another seven jointly for Cambridge and West Suffolk county councils.[21] They were very successful for a time and even such small villages as Barton and Dry Drayton had more than 50 students each. Although there was apparently little friction between the Syndicate and the County Council, the number of courses fell rapidly after 1894. Nevertheless the work continued and thirty years later the Council's village colleges began to take extension courses once more.

In order to carry out this additional work the Syndicate appointed a number of extra lecturers in science. At the peak of the demand it had two staff lecturers – C. W. Kimmins‡ for chemistry

* Cambs. Record Office, minute books of county and borough technical education committees, 1893–4. However as early as 1891 the Syndicate had organised three courses for the county (BEMS 22/2), while the borough had subsidised the work of the Cambridge Local Lectures Association (Borough minute book, 10 December 1891).

† Cambs. Record Office, borough minute book – *Report* of 17 November 1892. See also *University Extension Congress Report*, 1894, pp. 72, 73.

‡ Kimmins afterwards became Chief Inspector for the London Technical Board – the London County Council's technical education committee (B. Webb, *Our Partnership* (London, 1948), p. 80).

and E. A. Parkyn on hygiene and physiology – 22 other lecturers and three 'junior lecturers', but even this does not seem to have been sufficient.* In all the Syndicate directly assisted eleven local authorities by providing courses and at least the same number of councils approached the local Cambridge extension committee for help.† Almost all the work was done between 1891 and 1894 and during these years the number of centres rose from an average of 45 to 140, while the number of courses rose from 70 to 160. The numbers fell equally rapidly to 30 and 40 in 1895. The number of students did not show such a dramatic rise and fall and it is clear that the technical courses were on a different scale to the ordinary extension work. Although the Syndicate showed enthusiasm for the task in 1891, there were no expressions of regret in 1894 about the loss of this 'temporary work undertaken by the Syndicate'.[22] The *Annual Report* went on to regret that 'the cheap local technical classes organized independently by the local authorities' had adversely affected some Cambridge centres. It is clear that the councils increasingly felt that they could provide technical education more cheaply, while the Syndicate found the restrictions imposed both by the Act and by the councils increasingly irksome. In some places the Syndicate's courses were also too advanced and unsuitable for the students they were intended to attract. Many were in rural areas where the standard of elementary education was still low. At Tysoe courses in elementary mathematics and carpentry were successful, but 'lectures on architecture and astronomy, with lantern slides, and on the wonders of experimental science ... stirred no more than a little passing wonder'.‡ The inhabitants of Ightham in Kent complained that Mr Grünbaum's course on physiology was 'inappropriate'.[23]

The ultimate effects of the 'whiskey money' on University Extension were in the main unfortunate. Too much was demanded too quickly and the Syndicate became disenchanted with the scheme. The restriction to scientific and technical courses, which was maintained until 1903, was alien to the spirit of Univer-

---

* *Second Report on Instruction in Technical Education* (1892), pp. 18–20 (BEMS 22/2).

† E.g. Lowestoft and Wellingborough. These do not appear as 'whiskey money' courses in the Syndicate's records.

‡ M. K. Ashby, *Joseph Ashby of Tysoe* (Cambridge, 1961), p. 227. These may not have been Cambridge courses.

sity Extension,* but when local authorities began to provide their own courses extension societies were obliged to concentrate on arts courses almost entirely. At Cambridge in 1894 the 'old' University Extension Committee provided literary courses and the 'new' Technical Institute scientific courses,[24] although in this example the extension course was subsidised by having the free use of a room at the Institute. The emphasis on literary courses after 1894 tended to increase the attendance of the middle classes at the expense of the working classes. On the other side most local authorities became convinced that the Syndicate's courses were too expensive, and possibly too advanced. This belief persisted some forty years later and hindered later efforts at co-operation. In 1931 the clerk of the Peterborough Joint Education Authority summed up this attitude in his reply to the Cambridge Secretary:

Having due regard to the class of student and the nature of the classes for which there appears to be a demand the Committee are of the opinion that these classes can with equal effect be supplied by them and at a very considerably lower cost especially in so far as payments to lecturers are concerned. Should the Committee receive a sufficient number of applications from intending students they will be prepared to consider the provision and organisation of such classes under their own authority and supervision.†

When R. D. Roberts became Secretary he devoted much thought to the problem and in 1896 he drew up a memorandum for the Syndicate:

It is clear that from whatever point of view the Lectures work is regarded it is at present in a very critical condition. The number of Centres taking courses of lectures from Term to Term has been decreasing. The number of experienced lecturers free to take full work has been growing smaller. It is true that the lecture list is a long one, but almost all the senior and experienced lecturers have entered into permanent engagements which occupy part of their time at Cambridge, London or elsewhere. One reason undoubtedly for the decrease in the number of Centres is the want of first rate lecturers free to take work wherever the demand arises. There is another point which has been strongly impressed upon me during my visits to Centres during the past six months. At a large number of Centres the Local Committee have obtained aid from the Technical Education Fund of the County and Borough Councils towards courses on scientific

* Even though the restriction could be evaded for the most flimsy reasons. See B. Webb, *Our Partnership* (London, 1948), p. 80 and Derby Univ. Extension Soc., minute book 1, f. 8.

† BEMS 6/1, 270. The original enquiry was about the possibility of co-operation in rural areas.

subjects. As such aid is restricted to scientific subjects the tendency has been to neglect literature, history, etc. The tendency of these grants has unquestionably been to disorganise the Centres to some extent by enabling the work to be carried on with less effort, and so the local committees have tended to become dependent upon the County Council grants to such an extent that when the grant has been withdrawn the work has been dropped.[25]

It became Roberts' task to try to correct some of these mistakes and, in particular, to build up a new team of full-time lecturers.

One of the beneficial effects of the 'whiskey money' in which the Syndicate participated was its use to establish colleges in different parts of the country. During Berry's term of office the nucleus of the future University was founded at Exeter under the Syndicate's auspices. In 1875 and 1876 Exeter and several other Devon towns had Cambridge courses, but nothing more was done at Exeter until 1886 when an Extension Society was formed with a succession of enthusiastic secretaries.* In 1889 a reorganised society with Miss Jessie Montgomery as one of the secretaries applied for affiliation to Cambridge and at the same time took the initiative in forming a South Western Association for University Extension.[26] Miss Montgomery was a tireless worker for the extension movement both locally and nationally until her death in 1918.† She persuaded Exeter City Council to use its share of the 'whiskey money' to establish the Exeter Technical and University Extension College and persuaded the Syndicate to provide financial assistance for staff.[27] The nucleus of the college was the Albert Memorial Museum, School of Art and Free Library which an independent committee had erected in 1868 with voluntary contributions and a Treasury grant.‡ It provided art and science classes, evening classes and day technical classes. In 1891 the building was transferred to the city council despite some Treasury opposition,[28] so that the way was clear to establish the new

* Exeter University Archives, U.E.S. minute book, 1886–8. The first two secretaries, J. A. Hobson and H. Macan, signed the minutes instead of the chairman.

† *The Late Miss Montgomery*, n.d. [1919] (BEMS 32/1). A fund to her memory enabled Exeter students to go to Cambridge until 1940. It is now part of the Library endowment (BEMS 32).

‡ Public Record Office, Ed. 29/30. G. T. Donisthorpe, *An Account of the Origin and Progress of the Devon and Exeter Albert Memorial Museum* (Exeter, 1868). The original Victorian Gothic buildings still stand in Queen Street. At about the same time Oxford was organising a similar college at Reading with help from Christchurch.

college in 1893. A. W. Clayden, a Cambridge local lecturer, was the first Principal and for many years the Syndicate paid £150 of his salary for acting as superintendent lecturer in Devon.* The constitution of the College provided for a Museum and Technical Education Committee (with Technical, and Science and Art sub-committees) and a University Extension Committee.† In 1895 a local bequest enabled the building to be extended. The Chancellor of Cambridge University opened this and described the establishment of a university college in a smaller town as 'an experiment of very strong interest'.[29] Although many of the full-time students took a London external degree or a teacher's diploma, the links with the Syndicate remained strong. Until his retirement in April 1920 Clayden slowly led the college towards independence as the University of the South West.[30] His successor, Hector Hetherington (1920–5), placed more emphasis on association with Oxford, but the Syndicate still continued to have two representatives on the Court of University College.‡ The Syndicate continued to provide extension courses at Exeter until 1932, while University College was providing extension courses and tutorial classes in Plymouth and Cornwall from 1925 onwards.[31] This anomalous arrangement was the result of Exeter's anxiety to provide something for West Devon and Cornwall. It was not until 1945 that Cambridge withdrew from North Devon at the request of University College.§ It was then that the last links with Cambridge were broken.

The success of the Exeter experiment led to many other proposals to amalgamate extension work with technical colleges, but only the Colchester College came anywhere near success. Colchester had the Albert School of Science and Art in the Old Corn Exchange in the High Street, which was bought in 1893 with grants from the Treasury, Essex County Council and the borough

* BEMS 5/1, 36 *et seq.*; 5/2, 10 and 24. Local Examinations Syndicate, minute book 3, 11 March 1893. Clayden gave a long description of the College to the 1898 Conference (BEMS 28/3, pp. 42–8).

† *Calendar of Exeter . . . College*, 1893. Unfortunately most of the minutes and other records were destroyed when University College was bombed in World War 2.

‡ Hetherington, *University College at Exeter*, p. 15. BEMS 31/6 and 9. One representative was the Rev. J. H. B. Masterman, formerly a local lecturer and then Suffragan Bishop of Plymouth.

§ BEMS 37/20; 48/14 and 18. Exeter formed a Joint Extension Board with Bristol and Southampton in 1923, but had its own Dept. of Adult Education from about the same time.

council.[32] The very active Colchester University Extension Society immediately foresaw the possibilities.*

The Round family were the driving force behind the proposals. Mr Douglas Round made the original suggestion and Miss Juliana Round, as secretary of the Society, worked hard for its realisation.† They brought both Miss Montgomery and Mr Clayden to Colchester to describe what they were doing,‡ and the borough council established a Technical and University Extension College Committee to draw up a constitution. The Syndicate nominated one of its lecturers, Philip Lake, as first Principal and helped with his salary. The new college was opened by Lord Rosebery on 20 October 1896.§ It is not at all clear why it failed to develop along the same lines as Exeter, although the difficulties were undoubtedly great at both places. The altered arrangements for education after 1902 or the suspension of local extension courses during World War 1 may have been the cause, but by 1919 the two constituent parts had gone their separate ways. The records of the Extension Society show that the borough's Higher Education Committee made it a small grant and the courses were usually held in the Technical College – but nothing more.[33] The Board's archives are silent on the subject.‖

Other attempts along similar lines did not get so far. We have already seen how Cambridge borough council had brought the extension society into their Technical Institute, but this went no further. At Peterborough in January 1896 a public meeting discussed the advantages of an extension college, but nothing was done.[34] In 1911 the impending bankruptcy of the Plymouth University Extension Society led to a less ambitious amalgamation. An energetic borough Secretary for Education persuaded the town council to take over the running of the courses. There was an inaugural meeting in the town at which the Secretary of the Syndicate spoke, but there was apparently a clash of personalities and the courses soon came to an end.[35] Other attempts were made

* The records of the Society do not begin until 1919, but it was established between 1889 when courses began and 1895 (BEMS 33/2. *Essex Standard*, 21 September 1889).

† BEMS 5/1, 57. BEMS 33/2. Miss Round died in 1929 after 40 years service to the extension movement (*Cambridge Bulletin*, no. 6, p. 9).

‡ *Report of a Conference. . .*, 1898, pp. 47, 48 (BEMS 28/3). This includes a description of the College.

§ *Essex Standard*, 25 January, 5 September–24 October 1896. BEMS 33/2.

‖ The University of Essex at Colchester is of course a separate development.

at Scarborough, Southport, Leicester and the Potteries in the early years of the century, but little was achieved beyond the establishment of co-ordinating committees.*

In 1894 the Government established an indirect subsidy to extension courses when it began to encourage pupil teachers to attend them. Pupil teachers were the apprentices of elementary school headmasters and competed for Queen's Scholarships which allowed them to go to a training college.† From 1894 until 1906 it was possible to claim 60 marks towards the Scholarship for a sessional extension certificate in certain subjects:

    (i) A period of English Literature . . .
    (ii) A special aspect of British History . . .
    (iii) A special aspect of Geography . . .
    (iv) The elements of English Architecture . . .
    (v) Geology.
    (vi) Astronomy.
    (vii) Meteorology.‡

Since it was necessary for local School Boards to pay the pupil teachers' fees, the demand varied with the Board's enthusiasm. The first two centres where the Syndicate provided for pupil teachers were Cambridge and Leeds.§

Until 1898 they usually attended the ordinary courses where they happened to be suitable, but from then onwards it became more usual to organise pupil teacher centres in collaboration with the school boards (local education committees after 1902). Soon however there were complaints of the unfair advantage given to those pupil teachers who lived in areas where extension courses were available. In 1905 the Board of Education proposed to cancel the regulation and the Syndicate collected opinions about the value of the work.[36] Questionnaires were sent to all the Cambridge lecturers and examiners who had been connected with it. All

---

* *27th Annual Report*, 1900, pp. 6 and 7. The Oxford Delegacy was no more successful even though it had commissioned plans for an extension college (Mackinder and Sadler, *University Extension*, 1891, pp. 124, 125). BEMS 5/2, 118.

† Sue Bridehead in Thomas Hardy's *Jude the Obscure* (1896) was a pupil teacher who obtained a scholarship.

‡ *University Extension Journal*, vol. 1, p. 20. Cambridge introduced the sessional certificate in 1894 for the benefit of pupil teachers. Later the syllabus was amended to English Language and Literature, Geography, History or Languages.

§ *University Extension Journal*, vol. 1, p. 21. However some school boards had made local arrangements for the attendance of pupil teachers before 1894 (*Report of a Conference . . . 1890*, Cambridge, 1891, pp. 11–21).

described certain disadvantages – 'the Pupil Teacher who is both stupid and idle gets no advantage by being driven to attend such courses', and the need to amend the syllabus for their benefit. But with one exception they felt that these students gained something which they could not find elsewhere. The extension courses opened new vistas to those 'whose training is too often very narrow and whose instruction has been too much at the hands of teachers who know nothing except at third hand, who have, that is, well studied text-books founded on bigger books, which last were sometimes written by people who had studied for themselves on original lines'.* Despite all the arguments of the Syndicate and of the other universities, the Board of Education decided that after 1906 it would not accept sessional certificates.[37]

There had been demands for a government subsidy for extension work before the 'whiskey money' had been granted. In 1889 a committee of M.P.s, lecturers and local secretaries had been formed to urge this on the government. Mr Macan of Exeter had been one of the leaders in making an approach to the Chancellor of the Exchequer.[38] The Cambridge Syndicate, at that time still controlled by G. F. Browne, had issued a statement that it could neither support the approach nor provide alternative finance, but that it would be willing to administer a grant if it was made.[39] At a conference at Cambridge in July 1890 one discussion was devoted to this subject and there proved to be a considerable division of opinion. When a resolution drawn up by the committee was put to the conference it was too dark to count the votes. Before the vote was taken again the following morning it was announced that Browne had persuaded the sponsors to remove all reference to the Syndicate from their resolution. To this alteration there were further objections and eventually an amendment to seek the opinion of local centres was passed.[40]

Instead of proposing that county and county borough councils should extend their field to non-technical adult education, the Committee then decided to seek direct government aid. It was felt that the local authorities found it difficult to administer the 'whiskey money',[41] while there was a good precedent in the direct grant made to university colleges in 1889.† On 5 June 1891 the

* The quotations are taken from *Pupil Teacher Courses*, April 1905 (BEMS 37/8).
† The origin of the University Grants Committee.

Committee called a public meeting at the Westminster Palace Hotel at which Stuart, Sidgwick and Browne represented Cambridge. It was then decided to seek state aid not for the Syndicate, but for local committees or district associations.[42] However the impending fall of the Conservative Government seems to have delayed matters, and it was not until November 1893 that James Stuart led a deputation to the Chancellor of the Exchequer.[43] In January 1894 Arthur Berry persuaded the Syndicate to take the initiative and approach the Chancellor direct.* Unfortunately the only answer was that the Treasury had no more money to make available, and the Chancellor added the warning that grants 'may carry with them ultimate danger to the freedom from Government interference with their work'.† Although more efforts to secure a grant were made in subsequent years,[44] no further progress was made for nearly twenty years.

During his secretaryship Roberts also experimented with 'introductory lectures'. One of the difficulties of establishing new centres was to test the public demand. The Syndicate experimented with half courses from 1885 onwards, and individual lecturers gave single talks on their own initiative.[45] In 1888 the Gilchrist Trust entered the field of university extension by offering the London Society a grant for short introductory courses.‡ Roberts, then secretary at London and assistant secretary at Cambridge, persuaded them to offer a similar grant to the Syndicate.[46] This was the beginning of a long and happy association between Cambridge and the Trust. Both Roberts and his successor at Cambridge, Somerset Cranage, were secretaries of the Trust for long periods, and Gilchrist lectures were either organised through the Syndicate or in close collaboration with it.§ These courses either helped a centre which was in financial difficulties, or preceded the establishment of a new centre.‖ The grants were

* Local Examinations Syndicate, minute book 4, pp. 2 and 14.
† BEMS 38/12. The Syndicate felt that the danger was exaggerated.
‡ BEMS 53/12, 46/2, 42/1. The Gilchrist Educational Trust was established by a friend of George Birkbeck for general educational purposes. From 1868 onwards these included short courses of lectures for artisans.
§ See D. H. S. Cranage, *Not Only a Dean* (London, 1952), pp. 118–22 for a description of the Trust.
‖ In March 1896 the Gilchrist Local Committee at Newark decided to re-establish a Cambridge centre there (*University Extension Journal*, vol. 1, p. 110). The list, compiled in the Syndicate office, of extension courses (BEMS 26/1) illustrates this link by including Gilchrist lectures in each sequence.

usually made to suit local circumstances. In 1906, for example, the Bideford and Exeter centres applied for a grant to organise lectures in five Devon towns.*

In 1895, faced with falling numbers of centres and students, Roberts introduced a similar scheme of Pioneer Lectures. These were four weekly lectures by different lecturers on a common theme – usually history. They were offered by the Syndicate for an inclusive fee with a miniature syllabus.[47] This included a bibliography and four tickets for the four lectures. The local committee was therefore spared much expense. They were employed in exactly the same manner as the Gilchrist lectures and like them did not have examinations or count towards a certificate. Pioneer Lectures were obviously Roberts' idea. He had organised similar 'People's Lectures' for the London Society, and when he left Cambridge the Syndicate dropped them.[48]

Roberts' secretaryship also saw the establishment of a Local Lectures Library. The provision of textbooks for courses had been a problem from the beginning of university extension. The larger towns such as Leicester and Norwich had a borough library which might help, and smaller centres might build up a library of their own by gifts and subscriptions. The difficulty was to obtain a wide range of textbooks on different subjects, particularly since scientific textbooks were out of date so quickly.[49] In 1890 the Local Lectures Committee had begun to supplement local efforts by lending books from the Local Examinations Library.† This was not particularly satisfactory and in 1898, after a long correspondence in the *University Extension Journal*,‡ a Book Association for all universities' local centres was established. This predecessor of the National Central Library was the inspiration of Alice Thompson of Scarborough and other local secretaries.

While this was going on, Roberts had persuaded Miss Julia Kennedy of Cambridge to present a lending library which she had run for 27 years to the Local Lectures Committee to provide

* Exeter Archives Dept., town clerk's file B2/27. Later the Trust assisted the Workers' Educational Association and similar bodies.

† Local Examinations Syndicate, minute book 3, 8 June 1889. *Report on a Conference . . . 1890* (1891), p. 5. In 1895 books were lent for three months for one shilling (BEMS 5/1, 32). Oxford had 'travelling libraries' from 1885 (Mackinder and Sadler, *University Extension*, 1890, p. 79).

‡ It began in January 1897 with a description of the Bournemouth students' library (vol. 2, p. 52) and ended in July 1898 (vol. 3, p. 134). The Association was soon renamed the Book Union (vol. 4, p. 7).

a nucleus for its own library.[50] A sub-syndicate was appointed to make arrangements for its reception. Shelves were provided at the cost of Local Examinations, and the work of librarian assigned to the assistant secretary.* This library operated under very cramped conditions in the Syndicate Offices until Stuart House was built in 1927,† but it supplied book boxes to local centres for all courses.

In addition to his work for Cambridge and London, Roberts was also editor of the *University Extension Journal* and secretary of the Gilchrist Trust for long periods.‡ He organised an International Extension Congress in London in June 1894, and in 1898 a conference at Cambridge to celebrate the first quarter century of university extension. He emphasised the need for co-operation between universities and eventually brought the Joint Committee for University Extension into existence.§ He wrote widely about the extension movement – *The University Extension Scheme as the Basis of a System of National Education*,[51] *Eighteen Years of University Extension*,[52] and *University Extension under the Old and New Conditions*.[53] His chief concern was to emphasise the need for continuous courses of study followed by a diploma or a degree, and the need for a government subsidy if university extension was ever to achieve its full potential.[54]

In 1902, after the University of London had been reorganised as a teaching university, the functions of the London Society were taken over by the University.[55] It was probably inevitable that Roberts would be offered the post of Registrar of the University Extension Board. He had played an important part in getting the university reorganised with new statutes and had done much to build up the work of the Society. On 16 April Roberts wrote to the Vice-Chancellor of Cambridge explaining that 'new and important developments' at London had led him to accept, and he would leave Cambridge at the end of the academic year.[56] His resignation

* Local Examinations Syndicate, minute book 4, p. 231.
† Books were shelved in many of the offices and on landings, while book boxes were made up in the basement (information from Mr B. W. C. Green, former chief clerk). At one time the books were shelved from right to left (BEMS 4/3, 70). See also *University Extension Journal*, vol. 4, p. 66.
‡ BEMS 53/12; 46/2; 42/1.
§ This body founded in 1903 – and not the Central Advisory Committee on Tutorial Classes as stated in Kelly, *Adult Education*, p. 253 – was the first joint universities organisation. See BEMS 5/1, 22 and BEMS 38/15 for its minutes.

was accepted with much regret. At London, despite failing health and some difficulties with the newly-formed Workers' Educational Association, he was very successful.[57] He had retained his links with the University of Wales throughout his life and his last achievement was the organisation of extra-mural work in his own country.[58] In 1910 he persuaded the University to appoint a committee for the purpose, and in conjunction with Thomas Jones[59] he collected information and planned a Welsh University Extension Board. He died on 14 November 1911 before his plan was accepted – as it inevitably was. After James Stuart, R. D. Roberts was probably the most important figure in the Extension Movement.

To replace Roberts the Syndicate chose A. J. Grant, Professor of History at Leeds since 1897 but an extension lecturer of twenty years experience. Grant refused and a committee of the Syndicate recommended another former local lecturer, J. Holland Rose.[60] The Syndicate rejected the proposal and approached R. G. Moulton who had left the Syndicate to become Professor of Literature at Chicago in 1894.[61] Moulton refused by telegram and a split developed in the Syndicate between those who supported Holland Rose and those who preferred another local lecturer – D. H. S. Cranage of King's. After meeting on two consecutive days the Syndicate decided to advertise the post.[62] Four applications were received and two local lecturers were interviewed – G. E. Green (appointed in 1886) and D. H. S. Cranage (appointed in 1892).[63] Although some members of the Syndicate probably remembered Cranage's father's quarrel with the Syndicate in 1877,[64] he was appointed Local Lectures Secretary on 17 July 1902[65] – a post he was to hold for 26 years.

# 7

# D. H. S. Cranage, 1902-24

Three secretaries of Local Lectures were clergymen of the Church of England and all three later obtained high office in the Church. Two of them – G. F. Browne and D. H. S. Cranage – were in charge of the Syndicate for the greater part of its fifty-one years' existence.* Both were interested in medieval history and members of the Society of Antiquaries, but there were few other resemblances between them. Dr Cranage devoted most of his energies to the work of university extension and showed none of Browne's ability to amass offices.†

Somerset Cranage belonged to a minor Shropshire family which, at the time of his birth, lived at the Old Hall, Wellington. His grandfather, a local solicitor, had died at the early age of 34, leaving his widow to bring up their family by running a girls' school at the Old Hall.[1] At the age of 19 Somerset Cranage's father began a boys' school there. Because of ill-health and lack of money Joseph Edward Cranage was largely self-educated, but he obtained by examination a doctorate of Philosophy from Jena University. He continued to run the school until just before his death in 1891 and an obituary described it then 'as one of the most widely-known establishments for gentlemen's sons in the Midlands'. It was a progressive school, teaching German and 'Public Affairs' and having its own laboratory at a time when most schools still concentrated on the classics and mathematics.[2] His success as a schoolmaster is shown by the additions made to the Old Hall buildings and by the offers of posts elsewhere.[3]

Although he was never ordained in any church, Dr J. E. Cranage was deeply religious and in 1850 he founded Wellington

---

[1] For References see p. 202.

\* The third was the Rev. V. H. Stanton who was joint secretary with James Stuart for a year.

† Cranage's chief work beyond local lectures was to serve on the University's building and similar Syndicates. See Cranage, *Not Only a Dean*, pp. 147–51.

New Hall Mission where services of an evangelistic and ecumenical type were held for the poor.[4] He was in sympathy with all denominations except Roman Catholics and Unitarians, did not believe in infant baptism and was strongly attracted to the Millenarianism of the Plymouth Brethren. When his son entered King's College his tutor described the father as 'Dr Cranage, schoolmaster, not Rev., evangelical and blue riband'.* He was also concerned in the foundation of the Brotherhood and the Sisterhood – movements which still combine non-denominational services with adult education. In 1885 at a meeting in Dr Cranage's house a Mr Blackham of West Bromwich spoke of his success in organising a Pleasant Sunday Afternoon Class,[5] and from this talk sprang the two movements. Earlier still Dr Cranage had lectured locally for the Royal Society of Arts on science, and he had brought a lecturer from the South Kensington School of Cookery to talk on 'A Good Dinner and How to Cook it'.[6] He also helped to establish an Oxford extension centre at Wellington, which became a Cambridge centre in 1898 and closed from lack of funds before 1903.[7]

Somerset Cranage was his youngest son, born in 1866 and educated at the Old Hall School and King's College. Between school and university his father arranged for him to travel to India and Ceylon on a cargo boat, doing some work in exchange for his passage. At King's he read mathematics at his father's request although his interests were already elsewhere. A serious illness prevented him from obtaining honours and in 1890 he returned to Old Hall to teach. However in the following year he became a Cambridge local lecturer in history and architecture. He was already an experienced lecturer on these topics, having given talks to The Wrekin Society, and in 1888 he had bought a magic lantern to illustrate his lectures.[8] His course on the history of English architecture was very popular, but he was unable to accept all the requests from local centres because of continued ill-health.[9] In addition to illustrating his lectures, with slides, he arranged for the last lecture of his course to be held in a building of architectural importance in the neighbourhood. Even after he became Secretary of the Syndicate he continued to lecture on the subject, and in January 1926 he broadcast six talks on monastic buildings which were afterwards printed as *The Home of the Monk*.[10] Earlier he had

---

* *Annual Report of King's College Council* 1957–8, p. 23. 'Blue riband' means that he was a total abstainer from alcohol.

compiled *Churches of Shropshire*, a comprehensive work issued in parts 'at considerable financial loss'.[11] The latter obtained him the Cambridge degree of doctor of letters in 1913.

Over the years he had left behind his father's somewhat unusual religious views and in 1897 he applied for ordination to Bishop Percival of Hereford.* From 1897 to 1902 he was successively curate at Little Wenlock and Much Wenlock, where his widowed mother came to live with him. She moved with him to Cambridge, where she died in 1922:

Throughout these years, on any Sunday morning, a dignified don, conspicuous by his great bald head, and looking exactly like a monk, might be seen on his way to King's College Chapel, with a little old lady in black silk on his arm. These were Cranage and his mother: they were absolutely regular in their attendance.†

After her death he married Miss D. E. Tyrer of Bedfordshire.

All of Cranage's friends give the same account of him. He combined a surprising ability to get his own way with a capacity for making and keeping friends. Dr Scholfield, the University librarian, said he showed great ingenuity in devising ways of explaining something while giving the impression that his hearers knew it already.‡ He was capable of holding up the train to Royston (where he went to play golf) while the office boy brought his sandwiches from home.§ He was tireless in his efforts for adult education and medieval history, yet he managed to remain friends with all the rivals in both fields. G. G. Coulton (who was a Syndicate lecturer) and his Roman Catholic adversaries found Cranage a sympathetic listener.[12] He was liked by supporters of the Workers' Educational Association as well as supporters of university extension. In later years when he was Chairman of Convocation he dealt strictly but kindly with speakers who offended by declaring them 'contumacious'.‖ Since he was to encounter many

---

* Bishop Percival was a supporter of adult education. The Percival Guildhouse at Rugby is a memorial to him. See p. 156 below.

† *Annual Report of King's College Council*, 1957–8, p. 25. *Lucy Anne Cranage –* privately printed memoir, 1922.

‡ I owe this information to another of Dr Cranage's friends, Mr John Saltmarsh of King's.

§ This anecdote was given to me by the office boy, Mr B. W. C. Green, later chief clerk of the Board.

‖ He was elected Chairman of the House of Clergy in 1932 and Prolocutor of the Lower House of Convocation in 1936 and held both offices until 1945. The anecdote was given to me by Mr Saltmarsh. Contumacy was a medieval ecclesiastical penalty which Dr Cranage had presumably unearthed during his historical research.

difficulties and problems in his twenty-six years as Secretary, his character and enthusiasm were ideally suited for the tasks ahead.

His enthusiasm for the work was not as vocal as that of James Stuart or R. D. Roberts, but it was equally sincere. His chief contribution to the literature on the subject was his essay on 'The Purpose and Meaning of Adult Education',[13] which he said 'might perhaps be called a confession of faith'.[14] By the time that he wrote this essay, adult education had been accepted almost everywhere as an important and continuing element in British life. University extension was almost fifty years old and had justified its existence. His essay strikes just the right note for the period. He emphasises the need to help those who have failed to obtain education earlier in life, the continuing need for education throughout life, and the lowering of barriers brought about by an increase in education. The quiet tone in which it is written is reassuring – it bears out Dr Scholfield's remarks on his power to persuade. For a quarter of a century he used this power wisely. He enabled the Syndicate to survive several crises as well as World War 1. He saw the gradual introduction of government grants to support adult education. He presided over the establishment of the Board of Extra-Mural Studies and the building of its own offices. It was all done with that calmness which never seems to have failed him.

Soon after his appointment as Secretary, the Education Act of 1902 swept away the school boards and transferred increased duties to local councils.[15] The Act repealed the Technical Education Act and made it possible for the first time to support literary courses. A number of articles (one by Somerset Cranage) appeared in the *University Extension Journal*, reminding local centres of the possibilities and suggesting they seek representation on local education committees.[16] The Southport Society successfully applied to the Board of Education for a grant towards a course on the Napoleonic Era in November 1902. In March 1903 the secretary unsuccessfully applied for a seat on the local education committee.* By 1914 the Portsmouth and Derby societies had also sought grants from the Board of Education[17] and other centres had approached their local authorities. Unfortunately the suspicions

* Liverpool University Archives, Southport Univ. Ext. Soc., committee minute book, 1896–1918, ff. 29–32. The Act did not come into force until April 1903.

which had grown up earlier were responsible for some reluctance to apply for or to grant subsidies.

The 1902 Act also made possible the foundation of a new body for adult education – the Workers' Educational Association. The tutorial classes which it sponsored were rivals to the affiliation courses, and its local branches could be rivals to local extension societies.* This might well have caused difficulties, but at Cambridge there seems to have been little friction. The secretary and founder of the W.E.A., Albert Mansbridge, was welcomed there. He soon became a member of the Syndicate and in 1923 at the Jubilee Celebrations he received an honorary Cambridge doctorate.†

Albert Mansbridge was the son of a Gloucester carpenter whose family moved to Battersea when he was four. After leaving school at fourteen he worked first as a clerk in the Civil Service and then in the Co-operative Wholesale Society. He continued his education by attending evening classes at Battersea Polytechnic, extension courses at King's College London, and Oxford Summer Meetings.‡ He became convinced that the existing provision for adult education was inadequate. In 1898 he spoke to the Co-operative Congress at Peterborough on the need for adult education and in the following year to the Oxford Summer Meeting on 'Co-operation and Education in Citizenship'. His efforts to bring the two together were no more successful than Stuart's had been thirty years earlier.[18] In 1903 he returned to the attack with three articles in the *University Extension Journal*.[19] These were afterwards reprinted as a pamphlet – *Co-operation, Trade Unionism and University Extension*. This included an introduction by Holland Rose, a Cambridge extension lecturer, and an additional article by Robert Halstead. While recognising that James Stuart began his work for all classes and kinds of people, Mansbridge felt that the Extension Movement was now more concerned with middle-class students and middle-class subjects. We have already seen that this was partly true and

* Affiliation courses lasted for three or four years and followed a definite pattern of study. For more details of affiliation courses and university extension societies see chapter 8 below.

† In addition to the W.E.A. Mansbridge founded the World Association for Adult Education (1918), the Seafarers' Education Service (1919) and the British Institute of Adult Education (1921).

‡ This brief account of Mansbridge's life is based on his autobiography, the *Dictionary of National Biography* and M. Stocks' *The Workers' Educational Association* (London, 1953).

the result of financial difficulties and the reluctance of the working-classes to participate. Mansbridge's solution was very similar to Stuart's original proposals for artisans – a series of three year courses on economics, modern history and similar subjects of political importance.

Soon after the articles were published Mansbridge founded the Association to Promote the Higher Education of the Working Man, a title which he soon shortened to the Workers' Educational Association. In 1906 the Association in conjunction with Oxford University began the first tutorial classes at Rochdale and Longton (Staffordshire). The Board of Education provided a grant, originally 5s for each student, but later a block grant of £30 a year.[20] In 1909 the Association began similar classes in Cambridge local centres. Before that time the Syndicate had made clear its welcome for the new body. All the Cambridge secretaries, including G. F. Browne, sent donations either to the central fund or the local fund. James Stuart gave to both, and J. B. Paton sent a timely gift of £50.[21] In October 1903 Mansbridge asked for a Cambridge representative on the executive committee and the Syndicate nominated its secretary. Later this was changed to a representative on the advisory committee, and Somerset Cranage was transferred.[22] In February 1905 the Syndicate agreed to Mansbridge's request to become subscribers, and later in the year they also gave a donation of £25 from the Summer Meeting account. In 1906 they began the practice of awarding scholarships for W.E.A. members to attend Cambridge Summer Meetings.[23] In 1913 when the Cambridge University Association organised an appeal for funds, the Syndicate made its first priority: 'The most striking recent development of the work . . . the provision of Tutorial Classes for working men in connection with the Workers' Educational Association . . . It is desirable that at least £600 a year should be provided.'[24]

Most of the Cambridge local centres concerned welcomed this new development in the same generous fashion. At Derby and Southport the extension society offered a special ticket to all W.E.A. branch members, and at the former a W.E.A. representative served on the Society's council and *vice versa*.[25] Only at Portsmouth was there friction between the two bodies and that was a clash between strong personalities. J. Herbert Fisher, the secretary of the Extension Society, and E. T. Humby, the secretary of the W.E.A. branch, had strong opinions about adult education and found it impossible to collaborate.*

In April 1908 the W.E.A. sent a deputation to the Vice-Chancellor to suggest that Cambridge should begin to organise tutorial classes. By the end of the year the Local Lectures Committee were considering the proposal and in September 1909 classes began at Leicester, Portsmouth and Wellingborough.[26] Leicester and Portsmouth had active extension societies which continued to take courses from Cambridge. Wellingborough was a lapsed Cambridge centre and here the W.E.A. branch also organised extension courses for a time.† Wellingborough was also unusual because the class chose to study English Literature rather than the standard W.E.A. subject of industrial and economic history as Leicester and Portsmouth did.‡

The Syndicate and the Lectures Committee allowed Mansbridge a free hand in organising these classes. He recruited his own staff of tutors (subject only to Somerset Cranage's approval) and selected students in a completely different way to that followed by the extension movement. Their numbers were restricted to thirty which made for closer contacts between tutor and class, but prevented any possibility of the class being self-supporting.§ It also prevented other students from joining, although they might take the place of drop-outs later. Only a nominal charge was made to students and they were selected by the local W.E.A. branch without any advertising.‖ Although the W.E.A. branches advertised their single lectures and arranged for reports on them to appear in

* Another member of the Portsmouth W.E.A. with strong convictions was J. M. Mactavish, who afterwards became General Secretary of the W.E.A. (Information supplied by Sir James Matthews and Ald. J. Lacey, members of the first Portsmouth tutorial class.)

† *Kettering Leader*, 20 September 1912. The Wellingborough Extension Society was founded about 1933.

‡ The tutor at Leicester and Portsmouth was W. T. (later Lord) Layton (see Cranage, *Not Only a Dean*, p. 96). A. J. Wyatt was the Wellingborough tutor.

§ A similar proposal of small classes for 'the real students' had been made in 1894 (*University Extension Congress Report*, 1894, p. 48). The Edinburgh W.E.A., which afterwards left the Association, seems to have ignored the rule about numbers from the beginning (J. B. Barclay, *When Work is Done*, Edinburgh, 1971, p. 10).

‖ It is difficult to discover the precise arrangements for selection. There is some information in Parry (ed.), *Cambridge Essays on Adult Education*, pp. 146, 147 and in *Cambridge Bulletin*, no. 1, p. 17. Sir James Matthews believes that he only paid a subscription to the Portsmouth branch.

local newspapers, the early tutorial classes received no local publicity at all.\* When the homes of the students are plotted on contemporary street plans it can be seen how much their recruitment depended on personal contact. Almost all the students lived in the suburbs of the town in houses erected comparatively recently. They were all householders or the close relations (wives, sons or daughters) of householders. At Leicester in 1912 out of 31 students who can be identified only six were not in this category.†  At Portsmouth in the same year sixteen out of 26 were, and the rateable value of the houses they lived in averaged £14 which was comparatively high.‡ At Bedford between 1919 and 1921, 19 out of 34 students came within the householder category.§ The figures for Ipswich are particularly informative because classes in economics and literature were running at the same time –

Ipswich, 1919

|  | Economics | Literature |
|---|---|---|
| Total no. identified | 27 | 21 |
| Householders or relations | 14 | 15 |
| Average rateable value | £12 10s | £17 |
| Men | 24 | 8 |
| Women | 3 | 13[27] |

While this bears out Mansbridge's claim that the less well paid asked for classes with a political purpose, it is also clear that it was the subject rather than the type of course which affected the social status of the students. Extension courses in the same subjects would probably have attracted the same students.‖

The official statistics printed in the Tutorial Class *Annual Reports* unfortunately tend to conceal this information. Details of the students' occupations are not given separately for each class and there is a tendency to downgrade the occupation.¶ There is

* The newspapers of seven early Cambridge tutorial class centres – Bedford, Cambridge, Ipswich, Leicester, Northampton, Norwich and Wellingborough – have been searched without success. The Oxford Delegacy apparently insisted that no publicity should be given to their classes.
† The names of the students in BEMS 24/1 have been checked against the *Burgess List* for 1912–3 and *Wright's Directory* for 1914.
‡ The staff of the Portsmouth Record Office kindly checked the list in BEMS 24/3 against the local rate books.
§ BEMS 24/9 checked against the *Bedford Directory* for 1920–1. Unidentified students were usually those living outside the borough boundary.
‖ Unfortunately no extension course registers have survived amongst the Board's archives.
¶ E.g. 'shop assistant' includes the owner of a bookshop and 'insurance agent, etc.' a district manager.

also little indication of a high drop-out rate. Attendance at the first three classes is said to be 88 per cent in the first year and 86 per cent in the second,[28] but the surviving registers show that at Wellingborough in 1909 six students failed to do any written work and did not complete their first year. In 1912 nine members of the Leicester class resigned from lack of time and four because of difficulties with the work. At Portsmouth there were seven resignations in a class reduced to 26.[29] In most cases new students were recruited to fill the gaps and the best possible average was calculated for each student before striking a general average.*

It is not surprising that, although the Syndicate supported the establishment of tutorial classes to the maximum extent of its finances, it viewed the exaggerated claims of some of Mansbridge's supporters with suspicion. University Extension was already well-organised and supplying courses in a wider range of subjects to a wider range of people. At Leicester and Portsmouth the attendance at extension courses was much greater than that at tutorial classes:

Michaelmas 1908

|  | Extension Course | Tutorial Class |
|---|---|---|
| Leicester | 236 | 30 |
| Portsmouth | 73 | 30 |

Michaelmas 1909

|  | | |
|---|---|---|
| Leicester | 243 | 26 |
| Portsmouth | 70 | 32 |

Both Leicester and Portsmouth were affiliated centres, so that many of the extension students were following a three year course in exactly the same way as the W.E.A. students. In 1912 the Board of Education obtained a special report on tutorial classes and Leicester and Portsmouth were among the fourteen classes visited.[30] The inspectors noted the close resemblance to an extension course and, while impressed with the work being done, were careful to point out that it was not the exact equivalent of a university education.

It is unfortunate that the early achievements of the W.E.A. have been obscured by unsupported statements about its unique merits. It is clear that the 'workers' whom both Mansbridge and Stuart

* Although this is a fair method of calculation it produces figures higher than the methods now in use, and no direct comparison is possible.

attracted to their classes were not the workmen of *The Ragged Trousered Philanthropists*, who had neither the time nor the money to attend.* It was the teachers, the insurance agents, the railway clerks and the post office clerks in their 'villas'[31] who attended tutorial classes – just as they attended extension courses. The first class at Leicester was mainly composed of skilled artisans and managers living in recently-built houses,† but it also included the owners of a bookshop and a dairy, an insurance superintendent and the secretary of the Charity Organisation Society.‡ At Portsmouth the tutorial class students also attended extension courses,§ and at Wellingborough the W.E.A. branch organised extension courses.[32]

The W.E.A.'s claims aroused strong opposition in other areas. In 1919 a correspondent in the *Cambridge Chronicle* attacked an appeal for the University's tutorial classes fund, because he claimed that it was a front for Norman Angell, Bertrand Russell and the pacifists. A reply from the Syndicate that tutorial classes were 'non-political and non-sectarian' failed to end the attack.‖ The association between the W.E.A. and the Syndicate led in turn to attacks on the latter on political grounds. In 1932, for example, there was an attempt in the Senate to reduce or abolish the Board's grant:

> Since 1923, however, the general financial position has considerably changed for the worse, and it may be said with some probability that this deterioration is largely due to the activities, both in power and in opposition, of that particular party, which would professionally support the Extra-Mural cause. No doubt some members of the University may think that a sum of between five and six thousand pounds is a small price for keeping them placated . . .[33]

Despite this unpopularity the Syndicate and the Board have always supported the work of the W.E.A.

The special report on tutorial classes noted two interesting developments – preparatory classes and summer schools.[34] The

* Robert Tressall's novel was first published in 1914.

† My father, who moved to Leicester soon after World War 1, has confirmed my deductions from the printed evidence.

‡ BEMS 24/1. Fred. Satchell (bookshop), Horace Wells (dairy), Mark Peagam (insurance) and W. E. Hincks (C.O.S.). Another student, W. E. Wilford, was afterwards mayor of Leicester.

§ Information provided by Sir James Matthews who attended both.

‖ *Cambridge Chronicle*, 26 November – 31 December 1919. BEMS 6/1, 106. Norman Angell was alleged to have addressed a Cambridge Summer Meeting before the War, but I can find no evidence of this.

former were introduced by the W.E.A. as a two term course to test the demand for a tutorial class. The latter developed out of the special meetings for W.E.A. members at the Oxford and Cambridge Summer Meetings for extension students. In 1913 Cambridge organised a separate Summer School for fifteen tutorial class students.[35] In the following year and at the same time as the Summer Meeting there was another School with two courses on economic theory and two on English literature. Lodgings were found for all the students on Market Hill and the courses were given in the Arts Building nearby. The students were also admitted free to Summer Meeting lectures. From 1921 to 1939 the Summer School was held at Cheshunt College in its new building in Bateman Street.*

By 1913 it was clear that tutorial classes were a permanent part of the Syndicate's work. It was supported by the Board of Education's grants, individual subscriptions and annual grants from the more wealthy Cambridge colleges. At the end of 1913 the tutorial classes account even had more than £50 in hand. In 1911 Somerset Cranage had spoken at a Union debate in support of the work being done.[36] The Syndicate was now able to appoint a separate joint committee to carry on the work. This consisted of five members of the Syndicate and five members of the W.E.A. At its first meeting on 29 May 1913 Cranage and Mansbridge were elected joint secretaries.[37] From the beginning it conducted its business on similar and parallel lines to the Local Lectures Committee. It appointed its own tutors and organised its own trial lectures for those candidates about whom it was doubtful. The Summer Schools were largely independent of the Cambridge Summer Meetings. Its finances were kept separately. Although the maximum grant from the Board was raised to £45 in 1914, this was still insufficient to cover the cost and it was necessary to restrict the number of classes according to the funds available. Although it was said that the classes would eventually become self-supporting, it soon became clear to the Syndicate that this was most unlikely.

The Joint Oxford Committee which had recommended tutorial classes had also proposed 'that a proportion of the working-class students in such classes may pass regularly and easily' to a university.[38] The Cambridge joint committee adopted this suggestion

---

* Cheshunt College trains evangelical ministers – usually for the Congregational Church. It moved to Cambridge in 1906 and to Bateman St. in 1914.

and arranged for selected students to spend a year at Cambridge. The first two were chosen just before World War 1 by the Leicester and Portsmouth tutor.* Robert Law, a carpenter of Leicester, was given a sub-sizarship by Trinity College and a grant from Leicester Education Committee.[39] A young shipwright of Portsmouth, James Matthews, was also selected, but the local W.E.A. secretary, Mr Humby, objected to the selection procedure and was suspicious of the University's motives. The candidate withdrew to avoid a deep division in the W.E.A. branch and no one else was chosen.[40] Mr Law eventually spent two years at Trinity, but did not proceed to a degree. Some difficulty was experienced in finding him work afterwards, and the Committee had to promise financial assistance for a year. In May 1915 he became a clerk in the London Labour Exchange, but after the War he was replaced by an ex-service man, causing more problems. Later he lectured for the W.E.A.[41] Because of these problems and the effects of the War no further appointments were made until 1919 when a Birkenhead man was admitted to Fitzwilliam Hall.[42] In 1924 the scheme was incorporated in the wider plan for adult students' bursaries.† For some time there were W.E.A. suspicions of the scheme because it would remove the students from the working classes, while the University did not welcome the W.E.A. suggestion that these students should become W.E.A. lecturers on leaving university.

An activity which caused very few problems was the Summer Meeting, which probably reached the peak of its popularity under Dr Cranage. Like so many other ideas in University Extension the proposal to bring extension students to Cambridge during the Long Vacation goes back to James Stuart. He omitted all mention of this in his *Reminiscences* and speaking in 1892 he gave only the bare details: 'It was I think in the year 1884 that a sum of £10 was offered by a lady to enable the most successful student in the Examination of the next session to attend for a month at Cambridge. The sum was increased so that when the time came two pitmen and two young women school teachers were sent up. They spent a month at Cambridge and profited much from their sojourn.'[43]

---

* There is no indication that candidates from Wellingborough were considered – possibly because its subject was English Literature and not Economics.

† See p. 157 below.

R. D. Roberts, however, described the incident in some detail. Two Northumberland miners on a short visit to Cambridge in 1884 were greatly impressed by Trinity College. '"Oh! that it were possible," said one, "for some of our students to come up for a short time to work in Cambridge and see all this for themselves." The idea laid hold of the party. Why should it not be done? They were presently at tea in Prof. Stuart's rooms. The subject which filled their minds was soon mentioned . . .'[44] Miss Gladstone, who was present, enlisted the support of her father who supplied one scholarship,* and the other three were raised on Tyneside.

G. F. Browne was still Secretary at this time and in 1885 the visit of the students was arranged 'informally (the Syndicate being in no way a party to the scheme)' by R. D. Roberts.[45] The students were able to work at the Geological Museum and the Physiological Laboratory in continuation of the previous winter's lectures. It was so successful that Roberts recommended an enlarged scheme under the auspices of the Syndicate. This was not accepted, but two years later four scholarships enabled three miners and a schoolteacher to spend a month working on chemistry and physiology.[46] Little more was done until after Browne's resignation, and in the meantime Oxford had begun Summer Meetings on a much larger scale.†

In the summer of 1890 41 students spent a month in Cambridge for study and in 1891 47 – although not all were able to stay the whole time.[47] Students from Cambridge centres as far apart as Plymouth and Sunderland attended. They lived in Colleges, worked in the laboratories and attended scientific and literary courses. Although it was a great success the Syndicate considered that students tried to do too much and did not stay in Cambridge long enough.[48] In the following year Arthur Berry organised a more elaborate meeting to celebrate 21 years of University Extension and invited James Stuart to give an inaugural address.[49]

The first real Summer Meeting came in 1893 when the Syndicate took advantage of Oxford's abstention to organise a much larger assembly of 650 students.[50] The restriction that only Cambridge

* The W. E. Gladstone scholarship continued to be given as late as 1892 (BEMS 12/2, 303).
† Oxford got its inspiration not from Cambridge, but from the U.S.A. where similar meetings had been held at Chautauqua for some years (*University Extension Congress Report*, 1894, p. 28 – BEMS 2/27).

students could attend was lifted* and a much larger programme was arranged. Despite the Syndicate's misgivings it was a great success. The Regius Professor of Greek gave an inaugural lecture at the Guildhall on 'The Work of the Universities for the Nation',† G. F. Browne preached at Great St Mary's on Ecclesiasticus ix. 10,‡ and many dons gave lectures. There were excursions to neighbouring towns, a conference on local organisation, another on university extension for the working man, a Union debate and a conversazione. The Syndicate decided to hold similar meetings not more often than alternate years and 'classes on a smaller scale' in the other summers.[51] However in March 1894 they decided not to hold a meeting the following year 'on account of the probable condition of the drains'.[52] The 1896 Summer Meeting was organised by Roberts and the 467 students included '69 foreign students who came from France, Belgium, Germany, Austria, Denmark, Norway, Sweden and the United States of America'.[53] This marked the beginning of Cambridge's continuing association with courses for foreign students.

R. D. Roberts now reached an informal agreement with the Oxford Delegacy to hold summer meetings in alternate years only. In 1898 Cambridge refrained to allow the London Society to celebrate its majority with a meeting,[54] but from 1900 the pattern of summer meetings in the even years and vacation courses in the odd years was established. It continued with few alterations until World War 2. In 1904 the British Association met at Cambridge so the Summer Meeting was held at Exeter. In 1910 the Church Congress was at Cambridge so the Summer Meeting went to York.[55]

The Summer Meeting soon became an institution with an atmosphere of its own. Although a considerable number of scholarships for it were offered by the Syndicate, students' associations, societies and other bodies,§ the great difficulty for most students

---

* Even students unable to attend extension courses were eligible. Local Examinations Syndicate, minute book 3, 19 November 1892.

† It was published by the University Press – BEMS 15/2.

‡ 'Whatsoever thy hand findeth to do, do it with all thy might.' *Report of the Fourth Summer Meeting ... Reprinted from the Cambridge Chronicle* (1893) – BEMS 15/2.

§ Some extension societies provided scholarships for anyone in the district association (Colchester Univ. Ext. Soc., minute book 1929–41, f. 11). Some trade unions and co-operative societies assisted their members to attend (*Report on Fourth Summer Meeting*, 1893, p. 25. *Kindness, Beauty and Learning*, 1906, p. 5). For attempts to widen attendance see BEMS 34/7, f. 21.

was to spend a month, or even a fortnight, in Cambridge. So the same people attended both the Oxford and Cambridge meetings regularly because they had the time to go. Extension society secretaries were there – Miss Rigby of Southport, Miss Montgomery of Exeter, Miss Thompson of Scarborough and Herbert Fisher of Portsmouth – to attend the annual meeting of the Local Centres Union.* There were other regular attenders such as Albert Mansbridge and J. M. Mactavish, or the anonymous family of four co-operators.[56] Each year students wrote enthusiastic accounts of their experiences in the *University Extension Journal*, but they had much to be enthusiastic about. Public figures gave lectures there and so did some University lecturers who were no longer able to give extension courses in term time. The range of subjects and lecturers was restricted only by the topic chosen for each meeting. Mrs Fawcett spoke on the 'Social Progress of Women' in 1893; John Burns on industrial relations in 1906; G. P. Bailey on the 'Principles of Aerial Navigation' in 1912;† and Sir William Lever on 'Industrial Evolution and Co-partnership' in 1914.‡

There were also the diversions. In 1906 the Syndicate subsidised a performance of the *Messiah* and a play.[57] There were tours of the colleges and trips to Ely or Bury St Edmunds. The Cambridge museums and libraries were open for more informal visits. For the 1900 Meeting the Syndicate produced *Cycling Hints for the Neighbourhood of Cambridge* which included a list of agents for hiring cycles.[58] There were other temptations to cut lectures: 'we fall from grace – and lecture – one by one, two by two, batch by batch. What becomes of the graceless ones? All of us know how fatal are the temptations to which they succumb. Sometimes it is tennis, sometimes it is tea-parties; oftenest of all it is the river.'[59]

Truants could comfort themselves with the words of the Secretary – 'Even Mr Cranage says there is no use trying to do everything.'[60] All too soon the month came to an end. The final meeting in the Arts School in Bene't Street was reserved for the speeches,

* Local Centres Union, minute book, 1912–36. Miss Montgomery also acted as secretary of the Cambridge Summer Meeting Reception Committee until her death. She was succeeded by joint secretaries – Miss Thompson and Miss Clark of Royston – who quarrelled about their duties.

† He thought that aeroplanes would 'tend to make warfare impossible' (BEMS 15/10).

‡ He distributed copies of Lever Bros Co-partnership trust deed (BEMS 15/11).

votes of thanks, presentations, and for linking hands and singing 'das alte schottische Lied: Auld Lang Syne'.[61]

During Cranage's secretaryship the amount of extension work remained constant. There were about 40 centres and about 5000 students for most of the period, although it fell to about half during World War 1.* Unlike World War 2 this seems to have had little other effect on the work of the Syndicate. At first there were difficulties. Foreign students had either been unable to get to the Summer Meeting of 1914, or had been unable to return home again.[62] Centres had financial problems and the Syndicate made some concessions,[63] but there seems to have been no proposal to close down 'for the duration' as so many other cultural organisations did. The reduction in the number of centres seems to have been the result of extra work for students and the transfer of lecturers to war work.† In October 1914 the Syndicate agreed to organise lectures on the causes of the War, and in 1916 the theme of the Summer Meeting was Russia.[64] The theme for the 1918 Meeting was 'The War and Unity' – unity in the Church, unity between social classes, unity of the Empire and of all nations. It was a shorter meeting than usual, but the theme was one of great importance at the time, and the lectures were published.[65] The Y.M.C.A. organised lectures to the troops, and it was not until 1916 that the universities were invited to send representatives to the organising committee. Dr Cranage eventually became chairman of the Y.M.C.A. Universities Committee and some Cambridge lecturers gave talks in France, Italy and Greece, but the Syndicate was not involved officially.[66]

At the end of the War there was a revival in extension work which did little more at Cambridge than restore the pre-War position.‡ Tutorial classes had suffered more heavily from the War because few people had been able to make the three year pledge required and fewer had been able to keep it.[67] They showed a rapid expansion from two to ten in three years and continued to increase to 24 at the end of Dr Cranage's secretaryship. The chairman of the Tutorial Classes Committee from its foundation in 1913 was Dr T. C. Fitzpatrick, President of Queens', who did much to encourage this growth. As Vice-Chancellor he took a particular

* See Table 1.

† See BEMS 5/4, 159. In 1919 the Syndicate asked each lecturer to describe his war work and this information is on their files.

‡ Despite Dr Cranage's belief in 'a striking revival' (*Not Only a Dean*, p. 116), the post-War figures do not differ greatly from the pre-War. See Table 1.

interest in adult education and emphasised that both extension lectures and tutorial classes were 'an established and essential part of the *normal work* of a University'.[68]

In 1923 University Extension and the Cambridge Lectures Syndicate were both fifty years old. It was already certain that Local Lectures and Local Examinations were to be divided again and almost the last important decision taken by the Syndicate was to organise a Jubilee Meeting at Cambridge. The Rev. W. H. Draper, Master of the Temple and a local lecturer of many years standing, was commissioned to write a history.* A list of all the staff was also published.† The meeting opened on July 6 with a Congregation at which honorary doctorates were conferred on Sir Michael Sadler (formerly secretary of the Oxford Delegacy), R. G. Moulton and Albert Mansbridge. Honorary masters' degrees were received by a lecturer, G. P. Bailey, Herbert Fisher of Portsmouth and Alfred Cobham of Southport.[69] Congregation was followed by speeches. The first, by G. F. Browne, gave such a distorted view of the origins of local lectures, that the American who followed him was able to refer to 'Browne the founder of the Extension Movement'. It was unfortunate that neither Moore Ede nor V. H. Stanton (both of whom were present) was asked to give reminiscences of the first years.‡ On the four following days there were discussions on topics of general interest. It was perhaps inevitable that this repeated many of the discussions of earlier conferences. Problems of finance and subsidies still concerned local centres. Adequate recognition of certificates by the University was now less important. The training of extension lecturers had become less pressing. Looking to the future some speakers urged Cambridge to make more provision 'for selected extra-mural students to embark upon continuous courses of study'[70] and to undertake more work in rural areas. The Board of Extra-Mural Studies, which succeeded the Syndicate in the following year, was to promote both successfully in the next few years.

* W. H. Draper, *University Extension 1873–1923* (Cambridge, 1923). This is a general history of the extension movement drawn chiefly from secondary sources. It contains a good chronological table.

† *Cambridge University Local Lectures, Historical List of Secretaries and Lecturers* (Cambridge, 1923). This contains a number of errors, some of which had been corrected by 1927 when Stuart House was opened.

‡ *Jubilee Celebration of Local Lectures* (1923) pp. 7, 8. Three years later at the opening of Stuart House, Moore Ede gave a brief account of James Stuart (*Cambridge Bulletin*, no. 2, p. 32. See also no. 1, p. 6).

# 8
# Early Lecturers and Centres

The earliest lecturers were apparently chosen on a personal basis without any particular organisation. Neither R. G. Moulton nor Edward Carpenter, two of the first lecturers, say in their biographies how they were chosen, but Moulton was selected by Moore Ede.[1] When larger numbers of lecturers were needed an advertisement was inserted in the *University Reporter* seeking 'gentlemen' to give lectures.[2] A number of letters of application have survived amongst Stuart's correspondence,[3] but nothing is known of Browne's procedure for selecting lecturers except his belief that the Syndicate provided work for unemployed graduates.[4] Arthur Berry put the procedure on a proper basis by issuing a printed form to all enquirers asking for academic qualifications, titles of published works, list of courses offered and previous experience, and the names of referees.*

From time to time the Syndicate made its need for lecturers known. In June 1903 there was an advertisement in the *Appointments Gazette*, and in October 1905 there appeared an anonymous article on the merits of extension lecturing as a career.† Some Cambridge graduates were sent to the Secretary by the University Appointments Board, but most candidates either approached him direct or were recommended. In 1899 Moore Ede wrote to Roberts from Rome: 'I have come across a man here named Powys, a Corpus man, who would like to get Extension work. He has been lecturing each evening to a party connected with a girls' school . . .'[5] And so after a trial lecture J. Cowper Powys, the poet and novelist, was admitted to the supplementary list. J. B. Priestley was recommended by Quiller-Couch, and T. S. Eliot by Delisle

---

[1] For References see p. 204.
* The original form omitted a space for the date which had to be added by the Syndicate clerk at the foot of the foolscap sheet.
† BEMS 55/1 – Cramphorn. BEMS 38/16. The article was probably by Cranage.

Burns.[6] Applicants were asked to submit a syllabus of a course. They did this in varying lengths. In 1910 Dr Rosslyn Bruce submitted a complete set of lecture notes on 'Reunion with God', but T. S. Eliot contented himself with giving the titles of courses which he already gave for the Oxford Delegacy.[7]

If the information seemed satisfactory and the Syndicate did not already have enough lecturers on the subject, the candidate was interviewed by the Secretary 'to explain to them the nature of the work and the mode of procedure, and to form an opinion as to their capabilities'.[8] The Secretary also collected references. Some referees disclaimed all knowledge of the candidate's lecturing ability and others were hardly complimentary. Will Spens wrote of one candidate that he would not get a University post, but would be 'of considerable use for extension work'.[9] Occasionally professional jealousy intervened. St John Hope wrote of Harvey Bloom in 1903: 'The man in question believes himself to be an antiquary, but his knowledge is extremely limited, and what little original work he has done was so ill done that it were better not done at all.'* Other opinions were more useful to the Secretary because they concentrated on the candidate's ability to lecture.

After passing all these tests the candidate was then asked to give one or more trial lectures.† In Berry's day these were given at the Cambridge Training College to the students, a few pupil teachers and members of the Syndicate:

A lecturer is, as a rule, asked to give a course of three lectures either on consecutive Monday evenings at 8.30 or on consecutive Thursday afternoons at 5. At the end of each lecture the audience are invited to ask the lecturer questions . . . At the end of the first and third lectures one of the Secretaries . . . and the Principal of the Training College discuss the lecture privately with the lecturer.[10]

Later it was more usually one lecture at Homerton College or the New Chesterton Institute.‡ There were occasional complaints about the quality of the audience. William Cardwell complained that it was 'difficult to speak before girls of sixteen or thereabouts and yet imagine them mature men and women such as apparently form the usual audience at extension lectures'. When offered a further trial he asked that the official party should refrain 'from

* BEMS 55/9. Harvey Bloom's application was rejected.
† A few very distinguished lecturers were excused this ordeal.
‡ The latter still exists in Holland Street, Victoria Road, Cambridge.

continued talking and audible verbal comment'.[11] Percy Scholes objected that he did not have sufficient time to include musical illustrations, while J. E. Barkworth was told on his arrival that the girls knew nothing about music.[12] On the other hand the girls of Homerton sometimes complained about being used as guinea pigs – 'the lecture left a great deal to be desired'.[13] David Hardman chose Trollope as the subject of his trial lecture, not knowing that Dr Cranage was an enthusiast. He was disconcerted to find a relation of Trollope in his audience, and later to be interviewed by the Secretary and told he had mispronounced the word 'bass'.[14] Cranage himself gave a trial lecture in 1911 which he described as 'a severe ordeal, much more severe than I inflicted on others years afterwards', and during his first course at Lichfield the assistant secretary appeared to make a report on him.[15]

The notes made by the Secretary at trial lectures have sometimes been preserved. William Cardwell's lecture was 'overloaded with names and purely geographical information' and he talked 'down to the girls too much'.[16] The notes on J. B. Priestley begin ominously 'reading, accent...'[17] Another member of Syndicate described another candidate as 'an enthusiast – with a strong tendency to boredom'.[18] Occasionally the Secretary might attend a lecture which the candidate was giving elsewhere.*He at tended tripos lectures for the same reason: 'Lecture spoken but some reliance on notes. Rather academic in style and slightly monotonous in voice... elocution quite good and pronounces quite well. Subject matter interesting and exposition quite lucid. Plenty of confidence.'[19]

The reading of lectures and a different accent were always anathema to the Syndicate. C. F. G. Masterman apologised to Roberts for reading his lecture and tactfully explained that the Oxford Delegacy permitted this. He also asked for the name of a teacher of elocution as some other lecturers did.[20] In 1903 E. M. Forster's name was added to the supplementary list 'with the understanding that he should practise lecturing as much as possible, and, in particular, deliver some lectures under the supervision of the Syndicate in the May Term'.[21] Even members of the Syndicate occasionally felt that candidates 'received rather severe treatment'.[22]

* In 1919 Dr Cranage went to London to hear C. E. M. Joad and E. Allison Peers (BEMS 55/24).

Until 1893 the Syndicate only admitted men to the list of lecturers. Even James Stuart had been unwilling to admit women,[23] although it is fair to say that there can have been few qualified candidates and some prejudice against women lecturers in local centres.* Twenty years later conditions were very different. Women could take a degree at London and sit the degree examinations at Oxford and Cambridge. A number of women had experience of lecturing to university standards. So on 4 November 1893 the Syndicate admitted Ellen McArthur, lecturer in history at Girton, to their list of lecturers.† No objection was taken at that time, and in its next *Report* the Syndicate referred to 'the desirability of occasionally employing women as lecturers (particularly at centres where the large majority of the students are women)'.[24] Professor Marshall then wrote to the Vice-Chancellor complaining about the appointment. He not only claimed that women were unsuitable for lecturing, but that it would also have an undesirable effect on their characters.‡ The Syndicate spent two meetings composing a reply. This said that the Syndicate could not discuss the effect of lecturing on the lecturer, but would take every precaution to ensure that the students did not suffer. On the procedural problem:

It did not appear to the Syndicate when the question was first considered, that it was necessary to present a special report to the Senate on what seemed to them to be an administrative question within their powers and of comparatively small importance. By doing so they would probably have raised a controversy of a much wider scope, and it would have been difficult to avoid giving altogether a false idea of the importance of the step which the Syndicate proposed to take.[25]

The reply continued that the Syndicate did not consider it was precluded from appointing women, and invited Professor Marshall to raise it in the Senate if he thought otherwise.

Despite these brave words Miss McArthur was never asked to give a course of lectures, and no further woman was added to the Syndicate's list until 1918. Then the absence on military service of so many lecturers permitted the appointment and employment of Geraldine Jebb and Cecile Matheson without further controversy.[26] Meanwhile the Syndicate had prepared the ground by borrowing

* The first 'lady lecturer' at Derby was Mrs Charles Masefield in 1928 (Derby Univ. Ext. Soc., *Annual Report*, 1928–9).
† BEMS 5/1 contains no reference to this appointment.
‡ I have been unable to trace the original complaint (which may have been a flysheet). These details are taken from the reply.

female lecturers from the Oxford Delegacy where a less restrictive attitude prevailed.* The first of these occasions was in 1901 for a course on Italian Art.[27]

The Syndicate admitted about four or five lecturers to its list each year, but numbers fluctuated according to the demand for courses and the number of qualified candidates. The latter was not such an important consideration because at least four suitable applications were received for every one appointed, and they tried to have no more than one or two lecturers in each field. This was important as Hamilton Thompson pointed out to Cranage because two lecturers offering similar courses tended to clash.† The total number of lecturers available at any one time was probably about 30.‡ This included perhaps a dozen 'professional' extension lecturers who depended on the work for a living over a long period. The rest were 'temporaries' who either supplemented their earnings elsewhere by giving a few courses, or were waiting for a different post to become available.

Perhaps the most famous of the Cambridge professionals was Richard Green Moulton – the first of three very popular Cambridge lecturers on English Literature.§ Moulton was the son of a Methodist minister of considerable scholarship. After teaching and taking an external London degree, he was able to enter Christ's College at 26 with a scholarship in classics.[28] Immediately after taking his degree in 1874 he began to lecture for the Syndicate, and was so successful that he made his living by it for 20 years. He brought before his students the theories of Matthew Arnold, in particular that Literature was a 'Criticism of Life'.[29] His courses on Shakespeare, Faust, Ancient Drama, and (in later life) the Study of the Bible were very popular and attracted up to 1000 students.‖ There were complaints like that of D. W. Samways in February 1886 that Moulton had 'grossly overstated his average attendance' the previous term and the numbers in his class had

---

* It was usual for Cambridge and Oxford to borrow lecturers from each others lists when they had no suitable lecturer available.

† BEMS 55/27. In this case the lecturer offering similar courses was Cranage himself.

‡ See the list of 1898 in BEMS 38/11. Six years earlier with the stimulus of 'whiskey money' the number had been higher.

§ Moulton also lectured on Greek and Biblical literature, but his successors, Percy Babington and David Hardman, confined their courses to English Literature.

‖ Moulton's papers are deposited with the Board's archives – BEMS 3/1–155.

been 270 and not 350.[30] But the three centres of Coventry, Leicester and Northampton preferred Moulton to Samways who snubbed questioners.[31] In that term he lectured at Lancaster on Monday, Barrow on Tuesday, Preston on Wednesday and Blackburn on Thursday.[32] In the winter of 1891–2 he followed a more strenuous programme – Newcastle on Tyne on Monday, Middlesbrough on Tuesday, Paddington on Wednesday afternoon and Hackney in the evening, Egham on Thursday afternoon and the City in the evening, Russell Square on Friday afternoon and Lewisham in the evening, and Cheltenham on Saturday afternoon.[33] The books which he wrote were based on his lectures, but he also found time to write and lecture on the Extension Movement. In 1885 he put forward a plan to the Syndicate to enlarge the work and establish salaried lecturers, and he supported it with *The University Extension Movement.**

During the winter of 1890–1 he took a holiday in the United States where he was bombarded with invitations to lecture. While there he was persuaded to accept a post in the newly-founded University of Chicago where he was Professor of Literature from 1892 to 1919.[34] His reasons for doing so were twofold. First he wished to lecture on 'The Literary Study of the Bible' and the Syndicate was unwilling to allow laymen to lecture on religious topics.† Secondly he wished to marry the daughter of a Sheffield supporter of university extension.[35] Even the hard-working Moulton was unable to support a wife by extension lecturing. After retiring he returned to England where the Syndicate made amends by asking him to lecture on 'The Bible as World Literature'.[36] He was able to attend the Jubilee celebrations of the Syndicate, but died in the following year.[37]

Another professional was Hamilton Thompson, the medieval historian,‡ who was appointed in 1896 and gave three experimental courses on the Riviera in Lent term 1897 on Victorian Poets and Novelists.[38] In 1903 he listed all the courses he had given for the Syndicate since that time.[39] His income had been:

* The plan is in University Archives, C.U.R. 57.2, 130. The pamphlet (BEMS 3/115) was published by Bemrose of Derby – a firm which was closely associated with the Extension Movement.

† In October 1885 a lecturer at Chesterfield had offended three local clergymen merely by referring to Emerson's Unitarian beliefs (BEMS 37/2, 16).

‡ He was afterwards Professor of History at Leeds and wrote several standard books on the history of the English Church in the Middle Ages.

| 1897–8 | £125 | 1900–1 | £200 |
| 1898–9 | £150 | 1901–2 | £325 |
| 1899–1900 | £175 | 1902–3 | £300* |

He described this as a very comfortable income so long as he remained a bachelor.

Another medieval historian, G. G. Coulton, is an example of a 'temporary' lecturer. While teaching at Eastbourne at the beginning of the century he was accepted as a lecturer.† In his autobiography he describes what must have been a common hazard for extension lecturers in those days:

In the autumn of 1902 there came a sudden flood; I was engaged to lecture on 'Chaucer and his England' at three Devonshire towns,[40] with my headquarters at Exeter. This divided my week equally between south and west, for I was lecturing simultaneously at Eastbourne.[41] The journey was always touch and go: a before-breakfast train from Exeter brought me just in time for my three o'clock lecture at Eastbourne. Once, however, my otherwise irreproachable landlady at Exeter omitted to wake me, and as I opened my eyes my watch showed the exact moment of my train's departure! My lodgings were on Haldon Terrace, a good quarter of a mile from the station, but, fortunately, down hill. By equal luck, I had left my things ready packed, and in a kit bag; so there was no need formally to close it; only to grasp the two handles and away ... Again though the train was made up at Exeter itself, which generally means a punctual start, yet on this morning it was a few minutes late. Most fortunate of all, the engine driver was a real sportsman ...[42]

For his three courses of twelve lectures in the West, and six lectures at Eastbourne, he received £96 and his travelling expenses. This enabled him to spend a winter holiday in Switzerland.[43]

It will already have been noticed that the Syndicate employed (or on occasion refused to employ) many lecturers who were afterwards to become famous. No complete list of Cambridge lecturers has ever been compiled‡ so it is impossible to provide statistics. James Stuart proposed that young fellows of Cambridge colleges should spend several years at extension lecturing before returning to the University, but it is clear that this rarely happened. They

* BEMS 55/27 – letter of 2 March 1903. It should be noted that he gave long and short courses in differing proportions as well as repeated courses (i.e. afternoon and evening) for which payment was at different rates.

† Coulton, like a number of other Cambridge lecturers, had previously been a clergyman. He was a more controversial character than Hamilton Thompson, but this did not prevent the Syndicate from employing him.

‡ The *Historical List of Secretaries and Lecturers* (1923) is not complete and does not note whether those appointed ever lectured for the Syndicate.

left extension work to become professors in the new English and the old Scottish universities; they went to Australia, Canada, New Zealand and the United States; some obtained chairs at Oxford; but only four returned to Cambridge as professors.* It is difficult to find reasons for this. They were good lecturers and sound scholars.† Many of their books were based on the courses they gave.‡ R. G. Moulton deliberately lectured on the subject before writing *The Literary Study of the Bible*.[44] In other fields, one lecturer (Owen Seaman) became editor of *Punch*, some became inspectors of schools or directors of education, and not a few became members of Parliament. It can only be a source of regret that so few returned to Cambridge. Extension lecturers clearly felt cut off from University life since they were not in Cambridge for two of the three terms. The University too would have gained from contacts with the outside world. When Quiller-Couch (him-self an Oxford extension lecturer) was appointed Professor of English Literature at Cambridge in 1913, Lord Esher wrote to A. J. Balfour: 'You seem to be rather horrified at the idea that Sir Arthur Q should deliver himself of a sort of "advanced University extension lectures". . . . This is precisely what I hoped he would do . . .'§

The Cambridge extension lecturers formed their own organisation about 1880.‖ For most of its existence it was known as the Cambridge University Extension Lecturers' Union. It negotiated with the Syndicate, discussed problems of common interest and held an annual dinner with leading exponents of extension work as

* V. H. Stanton was Regius Professor of Divinity, J. E. Marr was Wood-wardian Professor of Geology, J. Holland Rose was Vere Harmsworth Professor of Naval History and T. B. Wood Drapers Professor of Agriculture. Of these only Holland Rose could be called a professional.

† Cunningham and Clapham wrote standard textbooks on economics, Pares on Russian history. W. W. Watts, Balfour Browne and J. J. H. Teall were famous scientists in their day.

‡ Hamilton Thompson's first two books – *The Ground Plan of the English Parish Church* and *The Historic Growth of the English Parish Church* – in the 'Cambridge Manuals of Science and Literature' were based on the courses he had given.

§ Churchill College Archives, *ESHR* 19/2, 20 May 1913. Quoted by permission of Lord Esher. See also F. Brittain, *Q* (1947), p. 124.

‖ Early in 1908 it was said to have been established 'nearly twenty years since' (*University Extension Bulletin*, Lent 1908, p. 12). The first annual dinner of the Union was held in 1891 (BEMS 36/7, 2). The minutes of a meeting between the Syndicate and the lecturers on 11 July 1890 may mark its formation (Local Examinations Syndicate, minute book 3, 11 July 1890).

its guests. From its beginning the Syndicate allowed it to elect two representatives on the Local Lectures Committee. The representatives reported to the Union at the annual meeting at Easter and the Union considered whether any action was necessary. In 1906 when the Lectures Committee, in conjunction with Oxford and London, decided 'to restrain lecturers from having their names on more than one University's list', the Union vigorously objected 'that one University does not provide enough work for a lecturer who is a specialist in a subject for which there is no great demand'.[45] Dr Cranage attended their meeting in 1925 to explain the different arrangements under the Board. His presence caused some embarrassment in the subsequent discussion, and in future years the Secretary always arrived part way through the meeting.[46] The Union persuaded the Board to stop paying lecturers quarterly in arrears.* It discussed cheap railway fares for lecturers, the provision of hospitality by local centres, and the establishment of tutorial classes.† Despite the small numbers involved,‡ the Union performed a very useful purpose in uniting the lecturers and making representations to the University. The Union came to an end about 1940, but its place was taken by the staff meetings which began in 1946.§

It is unfortunate that we are as badly informed about the Cambridge students as we are well informed about the lecturers. This is partly because the Syndicate had little direct contact with students. No registers of students have been preserved earlier than 1909 and only a handful pre-date 1960.[47] For descriptions of the students we have to rely on quotations from secretaries and lecturers. R. G. Moulton summarised a number in 1885.[48] He quotes an examination list for an unnamed town: '31 were men, and 27 women; of the men 4 were students, 5 artisans, 4 warehousemen, 9 clerks and shopkeepers, 6 large manufacturers, 1 schoolmaster, 2 unknown; of the women, 7 were daughters of manufacturers, 2 of ministers, 12 of tradesmen, and 6 were of the

---

* Information provided by Mr Hardman who was secretary of the Union from 1932.

† BEMS 36/7, 7, 32, 101. Until 1939 local centres usually provided hospitality in members' homes.

‡ In 1909 34 people attended the annual dinner, but the number was usually less than 20 (BEMS 36/7, 2).

§ BEMS 11/1. The Union did not hold any meetings after May 1940, but was not formally dissolved until after the War (information from Mr Hardman).

milliner class'.* Moulton implies that all the classes of people mentioned by Stuart in 1872 were attending the courses – schoolteachers, young people between leaving school and getting married, men in business, ladies who 'had not read a sensible book' since leaving school and artisans, is how he describes them. He quotes extensively from Roberts' reports to the Syndicate for accounts of workmen attending courses, particularly in Northumberland and Durham:

Two pitmen, brothers, who lived at a village five miles from one of the lecture centres . . . They were able to get in by train, but the return service was inconvenient, and they were compelled to walk home. They did this for three months on dark nights, over wretchedly bad roads, and in all kinds of weather. On one occasion they returned in a severe storm, when the roads were so flooded that they lost their way, and got up to their waists in water.†

The fragmentary information available suggests that there has been little change over the years in the quality of the students – only in the quantity. Even if the figures given for R. G. Moulton's lectures are exaggerated, there are statistics for Liverpool courses which can be trusted.‡ In Michaelmas term 1879 Moore Ede began his course with 279 students and ended it with 113. Nowadays few courses attract as many as 50 students,§ but in this Cambridge does not noticeably differ from other universities.

Although a number of students of literature subsequently wrote books, few have chosen to describe their experiences at extramural courses. V. S. Pritchett attended a class at Bromley about 1918: 'The lecturer was a young woman with rimless glasses, an icy, cutting and donnish voice, who looked as pink as if she had just got out of a cold bath. She sliced the air above her respectful audience . . . we were firmly told to submit papers . . .'‖ A Cambridge student was once moved to write verse about his lecturer:

* R. G. Moulton, *University Extension Movement* (1885), p. 15. It is possible that the town was Nottingham, and the courses literature and political economy which were given in Michaelmas 1883.

† *Ibid.* p. 19. R. D. Roberts' *Report to the Syndicate*, 1883/4, p. 4 – BEMS 22/1, f. 93. See also Parry (ed.), *Cambridge Essays on Adult Education*, p. 173.

‡ I am indebted to Miss A. M. Platt for pointing out that the statistics for Wellingborough given by the Board differ from those given by the local Extension Society, showing that it is still difficult to calculate numbers.

§ E.g. *46th Annual Report*, 1970 lists one course with 50 and one with 60 students.

‖ V. S. Pritchett, *A Cab at the Door* (London, 1968), p. 213. Bromley was not a Cambridge centre and I have been unable to identify the course.

We are but little students meek
Not born to any high debate
How can we please our Tutors who
Are very highbrow, good and great.

Each week the Tutor takes his place
His mind a blank – his brain a fog
He gets his inspiration from
The student who recites the log.

And sometimes when we've just begun
To think our tutor quite humane
He starts to ask for essay work
And then we know he's quite insane.*

Another Cambridge student, Miss H. S. Cheetham, who was a student at Southport from the very first pre-Syndicate course of F. W. H. Myers in 1870, wrote in more serious vein:

To many an ambitious girl, touched with the enthusiasm of the new movement, whose heart ached with hopeless desire for a University career, and whom stern Circumstance held in a groove of hard work and soul-breaking drudgery, University Extension came as a gift from the gods. To some of us who felt the first glow of this New Learning for Women, the change from the old system of study to the new was like stepping out of a dark and rocky gorge of the difficult hill of knowledge on to a wide and sunlit upland . . . We defied the laws of health gaily: after burning up much midnight oil, we would get up at impossible hours to do our paper-work. We hurried over our meals . . . to get back to the beloved books.†

So long as the Syndicate was satisfied that there was a competent local organisation, it did not for many years concern itself with the administration of local centres.‡ It is sometimes impossible to discover from its records how a centre was being run. It was not until 1896 that the Syndicate showed any preference for a particular way of organising a centre, and only the previous year that it had approved a national organisation of centres. Since that time Cambridge has offered advice and assistance without trying to dictate local arrangements.

* Probably written between the two Wars about a tutorial class, the typescript of this poem was lent to me by Mr B. W. C. Green.
† Article signed 'H.S.C.' in *University Extension Bulletin*, Lent 1909, p. 11. Miss Cheetham was a founder member of the Southport Society in 1896 (Liverpool Univ. Archives, Southport Univ. Ext. Soc. sub-committee minute book, 1896).
‡ A French writer favoured this local autonomy regarding it as a 'trait charactéristique de la race anglaise'. (A. Espinias, *L'Extension des Universités*, Paris, 1892, p. 28.)

For the first four years almost all centres were run by a local committee usually, but not invariably, elected at a public meeting. Derby, for example had a 'Provisional Committee' of 8 members of the borough council, 6 clergymen, 3 J.P.s, 7 'esquires', 3 ladies and 10 'gentlemen'.[49] Norwich (1877) and Sheffield (1874) had similar local committees,[50] and they continued to be formed as late as 1887. In that year the Cambridge local centre began:

A small provisional committee has been formed, and the kind co-operation of the Mayor as president during his term of office, and of Prof. Westcott, Prof. Hughes, Mr W. Bond, Mr R. Bowes, the Rev. G. F. Browne, Mr J. E. Foster and Mr J. O. Paine as vice-presidents, has been promised in the event of an organisation being formed to carry out the scheme . . .[51]

In the previous year R. G. Moulton had addressed a public meeting at Exeter. The 50 people present elected themselves 'together with the members of the Cambridge Local Examinations Committee' as the first extension committee 'with power to add to their number'.[52] Here the large size of the Committee made it necessary to appoint an executive committee,* and the main Committee eventually became the Extension Society.

For a time there was an experiment in running local centres through literary and philosophical societies. In 1877 the Hull Society took over the work of the local committee, and its example was followed in 1882 by the Leicester and Newcastle upon Tyne Societies and in 1885 by the Southport Society.† All the centres except Leicester continued to be run in this way for a number of years. Much later, in 1945, the local centre took over the Lowestoft Literary and Scientific Association,‡ but this was unusual.

It was also in 1877 that the first two Extension Students Associations were formed at Sheffield and Hull. Others were formed at Nottingham in 1878, Derby in 1880, Scarborough in 1887 and Norwich in 1890.[53] Few seem to have survived for long periods and very few records have survived.§ Only the Derby University Students' Association printed an annual report.[54] Its purpose was

* This was also the arrangement at Norwich.
† *University Extension Journal*, vol. 3, p. 99 and vol. 5, p. 35. Leicester Archives Dept., Lit. and Phil. Soc. minute book, 24 July 1882. Liverpool Univ. Archives, Southport Univ. Ext. Soc., list of courses compiled in 1896.
‡ The Association organised Cambridge courses from 1936. From 1945 it confined its activities almost entirely to this (Association minute book, 1923–61).
§ The account book of the Scarborough Association is in Scarborough Public Library.

'to encourage and maintain an interest in the University Extension Scheme', and all past and present students were eligible for membership on payment of two shillings a year. It organised additional lectures and meetings,* and different sections tried to continue the work done in courses.† From 1883 for a few years its work was supplemented by an Artisans' Association for the Extension of University-teaching and of Higher Education.[55]

Meanwhile in 1879 a new form of organisation appeared in Derby, Scarborough and Hull in the form of the University Extension Society.‡ The Derby Society was composed of subscribers of at least one guinea a year and donors of at least ten guineas. In return they received tickets for the courses and the privilege of deciding their subjects. The Derby Society was governed by a council on which both the Students' Association and the Artisans' Society were represented.§ Very few societies on this pattern were formed for the next fifteen years,‖ but in 1896 and 1898 Somerset Cranage wrote two articles advocating their formation which the Syndicate thought worth reprinting and circulating as a leaflet.[56] In the following year the Syndicate also printed and circulated a draft constitution for an extension society. On behalf of the Syndicate Cranage visited local centres to talk on the formation of societies, which he described as a painless way of raising a guarantee fund.

A complete record of the formation of the Southport and Birkdale Society in 1896 has survived. Southport already had a students' association and, when the Literary and Philosophical Society wished to give up the local centre, a small group of students established a committee on 11 June 1896 to work for a Society.¶ They drafted a constitution on the usual lines and invited representatives of the local authorities, the Literary and Philosophical Society, trade unions, and the Co-operative Society to send representatives to the Council.[57] A public meeting was called for

---

\* In 1889 it had two lectures by Mary Bateson the historian.

† For similar arrangements at Backworth (Northumberland) see BEMS 21/1, 99.

‡ BEMS 22/1, 92. *University Extension Journal*, vol. 2, p. 99 and vol. 3, p. 99. *Annual Report of the Derby Society . . . Fifteenth Session*, 1894–5. The records of the Scarborough Society are in Scarborough Public Library.

§ *Annual Report* of the Society, June 1885, p. 3. The Society's minutes do not begin until September 1888.

‖ Colchester about 1889 seems to be the only one.

¶ Liverpool Univ. Archives, Southport Univ. Ext. Soc. sub-committee minute book, f. 1.

11 December immediately after a lecture to the Students' Association by Mr Sadler.* The new society was voted into existence with
his support. The annual subscription was 15s, but 'Assistant
Teachers, Governesses, Shop Assistants and Clerks' paid 8s, and
'Artizans and Pupils and Pupil-teachers' paid 5s.† Both the Derby
and Southport Societies still exist, and in 1970 there were 19
societies in East Anglia organising Cambridge courses.‡

The first attempt to bring local centres into association was
James Stuart's Sheffield conference in 1875, but for the next 20
years the Syndicate did very little to bring them together. About
1880 the mining centres of Northumberland and Durham (which
were particularly subject to financial crises) formed a joint committee under the auspices of the mining union. This committee
organised conferences and encouraged co-operation.[58] Seven years
later regional associations of Oxford and Cambridge centres began
to appear. The first seems to have been a North Midlands University Extension Society which was subsidising local centres in
1886.[59] In the following year Northern and South-Eastern Associations for the Extension of University Teaching were formed. The
Northern Association succeeded the miners' committee; it pro-
moted an annual conference and 'the grouping of centres for
lectures'.[60] The South-Eastern Association, as we have already
noticed, was very active in establishing new centres and in general
missionary activities.[61] In 1889 a South-Western Association based
on Exeter was established and about 1891 an Eastern Counties and
a Lancashire and Cheshire Association.[62] It is probable that the
availability of the 'whiskey money' stimulated the growth of these
Associations and when it was withdrawn they declined. Nothing
is known of their later activities.

With the holding of regular summer meetings it became possible
to have an organisation restricted to Cambridge centres. In
February 1895 the Cambridge Local Secretaries' Union was
formed with the encouragement of the Syndicate to discuss
matters of common interest and make representations.[63] They met
each year at the Summer Meeting. A small subscription was

* M. E. Sadler, then one of H.M. Inspectors of Schools and previously
Secretary of the Oxford Delegacy.
† *Ibid.*, f. 10. It is interesting to note that when the subscription was raised
to 17s 6d in 1920 that for 'artizans' was lowered to 4s (*Annual Report of Society*,
1920).
‡ Board of Extra-Mural Studies, List of Local Secretaries, 1970.

collected from each centre and a fund to send students to the Summer Meetings was established. At the fifth annual meeting of the Union on 4 August 1900 proposals were made (possibly by R. D. Roberts) for a Local Centres Union based on district associations.* By these proposals England and Wales was divided into eight districts and the centres advised to form an executive committee for each district. However the districts did not form the Union – the centres joined it and paid subscriptions directly to it. This was probably the result of the Union being formed before the districts had been established, but it later caused occasional disputes. The proposals were accepted at the annual meeting in August 1901 and the Northern and Eastern Districts were established almost immediately.† Bernard Pares, then a full-time lecturer for the Syndicate, undertook the task of getting the districts organised, and while doing so assembled useful historical notes about the centres.[64] Midland, Southern, North-Western and South-Western Districts were formed in the next few years. All the districts disappeared before 1945 with the gradual withdrawal of the Syndicate. The Southern District's minute book ends in 1930, the Midland District's by 1932, the North-Western District had ceased to exist by 1934, and the Eastern District held no meetings after 1940.‡ Because the latter became almost coterminous with the Union, its work was taken over by the Union. The latter has continued to flourish and now holds its annual meetings at Madingley Hall.

The type of course which James Stuart had devised continued to be given up to 1945. It was known as a terminal or extension course, because it had been joined by pioneer and short courses by 1897, by tutorial classes in 1909, and by rural area courses by 1927. The terminal course in 1939 still consisted of 10 or 12 lectures of one hour, followed or preceded by 10 or 12 one hour classes, which were not attended by all those who went to the lectures. An even smaller number submitted weekly papers and still less took the examination set at the end of term. But gradually over the years the distinction between lecture and class had been eroded and the

---

* A similar Oxford Local Centres Union was formed in 1901 (*University Extension Journal*, vol. 8, p. 6).

† BEMS 34/6 and 35/1. *29th Annual Report*, 1901. Local Centres Union, E. Anglian minute book.

‡ BEMS 34/7 and 8. Local Centres Union, minute book, 1912–36, f. 49. E. Anglian minute book with L.C.U. records.

number of students taking the examination steadily fell.* From time to time there were attempts to revive the practice of writing weekly papers. In 1946 the Board ceased to report the numbers taking terminal examinations, and in 1961 the numbers of those writing weekly papers. The distinction between lecture and class was finally abandoned in 1957, but had almost disappeared several years before. During the first fifty years the demand for scientific subjects declined steadily and was replaced by Literature, Art and Architecture.†

We know a great deal about the lectures which were given from a variety of sources. Even as late as 1935 an active local secretary could persuade the local newspaper to print several hundred words on each lecture.‡ Sometimes the lecturer's own notes have been preserved and from 1885 the Board has an almost complete set of syllabuses.§ Some of the early lecturers had their syllabuses printed privately by a local printer and sold them direct to the students. W. W. Watts had his early syllabuses printed by Leach and Son of Wisbech.‖ D. W. Samways had an elaborate syllabus on 'Physiology and Health' printed in London in 1891 for use in any centre where he lectured.[65] But from 1890 onwards the greater number were printed by the University Press at the expense of the Syndicate and with a standard format. The inside of the stiff paper cover had the 'Method of Conducting Lectures and Classes' at the front, and advertisements for Cambridge books at the back. It contained the questions or subjects for the written work and the entry form for the final examination. Lecturers were encouraged to use the same syllabus for as many courses as possible.¶ It was not until 1939 that the Board abandoned its practice of printing syllabuses.

* See Table 1.
† See Table 6.
‡ E.g. *Bury Free Press*, 28 September 1935. In 1909 the Southport U.E. Soc. threatened to withdraw its advertising from a local paper if it did not print longer reports (Liverpool Univ. Archives, Southport U.E.S. committee minute book, 1896–1918, ff. 8 and 66). In 1911 the Norwich U.E. Soc. complained that verbatim reports were keeping people away from their lectures (Norwich Record Office, Committee minute book, 16 March 1911).
§ Edward Carpenter's notes, printed lectures and syllabuses are in Sheffield Archives Dept. (MS. 12 & C374, 65); R. G. Moulton's are with the Board's archives (BEMS 3). Other syllabuses are in BEMS 17.
‖ Two bound volumes are in the possession of his daughter, Mrs Fearnsides.
¶ A description of how to compile a syllabus can be found in R. G. Moulton's *A Lecturer's Notes on the Working of University Extension* (Philadelphia, 1890) – BEMS 3/118.

A few interesting sidelights on the methods of lecturing have been preserved. E. S. Thompson lecturing at Plymouth in 1875 on English Constitutional History impressed the local reporter with 'a diagram of a most ingenious character, and one well calculated to produce discussion'.[66] Lantern slides were introduced quite early and the Board's collection still contains some ingenious moving ones. Somerset Cranage illustrated his lectures, and worked the magic lantern himself.[67] Lecture notes with copies of diagrams and other aids have been preserved for R. G. Moulton, Edward Carpenter and W. W. Watts. The latter also apparently enlivened his lectures with light verse:

### Torricelli

Seated one day at his labours,
He was weary and ill at ease
And his hand it ached with pouring
Over philosophy's keys;
He could not have told the reason,
But lo – to his mind 'twas clear
That pumping raised up water
By the weight of the atmosphere.*

Few of the weekly papers submitted by the students have survived,† but we know the subjects set and have some of the lecturers' reports on them in local centre records. Most were impressed by the standard reached by their students:

The weekly papers have been generally well and carefully prepared, and have shewn in some cases evidence of considerable reading in various histories . . .
The work done in the weekly papers has reached a high, in some cases a very high standard . . .
Several of the papers shewed that the men had been trying experiments for themselves . . .[68]

At Southport in 1882 W. W. Watts had 90 students at his lectures and 50 at his classes. Forty-five of the latter wrote at least one paper for him.

| No. of Papers | 10 | 9 | 8 | 6 | 5 | 4 | 3 | 2 | 1 |
|---|---|---|---|---|---|---|---|---|---|
| No. of Students | 3 | 12 | 7 | 4 | 1 | 2 | 4 | 7 | 5‡ |

* *Syllabus of Lectures on Physical Geography* (Scarborough, 1882) – Mrs Fearnside's copy.
† The *University Extension Journal* occasionally printed some of the best papers.
‡ BEMS 22/1, f. 87. *Syllabus of Course on Geology*, 1882 – Mrs Fearnside's copy.

Very little information can be found in the Board's records about the terminal examinations. A very small number of examination papers and the names of the successful candidates form the whole. Since the Local Lectures Committee conducted its own examinations independently of the Local Examinations, nothing can be found in the archives of the Local Examinations Syndicate either. Once again the most helpful source is the papers of individual lecturers and examiners. W. W. Watts preserved the printed *Directions to Students*, probably issued by Arthur Berry:

1. Be at your seat in the Examination Room five minutes before the time fixed for the Examination to commence.
2. Write your full name in the right-hand top corner of every sheet of paper which you use, adding your address on the first sheet.
3. Write only on one side of the paper. Write the number of each question before the answer. Begin each answer on a fresh sheet . . .*

A volume of Edward Carpenter's examination papers has been preserved with his archives.[69] He apparently set his own papers and had them printed locally. At first he set from 10 to 12 questions and allowed candidates to attempt them all in $2\frac{1}{2}$ hours. By 1877 he was restricting them to three-quarters of the questions set and varying the time allowed.

The instructions issued for examiners by Arthur Berry have been preserved by Oscar Browning:

The maximum number of marks is 100. This number of marks is to be assigned to the number of questions to which candidates are instructed to confine themselves. Thus if thirteen questions are set and Candidates are only allowed to answer nine, the full marks for any nine should be 100.

Those who receive less than 25 are to be rejected. The standard for distinction is left to the discretion of the Examiner, within the limits of 65 and 70 marks; in subjects of special difficulty, the minimum may be 60. Only those candidates who are recommended by the Lecturer on the result of the weekly work, as well as by the Examiner on the result of the Examination, obtain the mark of distinction in the final list.

If two or more places are taking the same course, a separate report should be written for each.[70]

Oscar Browning was a frequent examiner for the Syndicate, even though his career as a local lecturer was very brief. He had an unfortunate habit of leaving the examination room for long

---

* Bound into a volume of syllabuses, 1889–94, in the possession of Mrs Fearnside. Similar instructions appear in *Instructions for Local Secretaries*, n.d. (BEMS 38/9).

intervals which distressed G. F. Browne.[71] There were also disputes between lecturer and examiner about the merits of particular
candidates. R. G. Moulton had strong feelings on this subject.[72]
However few candidates who entered for the examination actually
failed:

|  | December 1877 | | |
|  | Class I | Class II | Rejected |
| --- | --- | --- | --- |
| Chesterfield | 5 | 10 | 0 |
| York | 49(7) | 10(3) | 4(2) |
| Sheffield | 8 | 17 | 5* |

From the beginning James Stuart had intended that students
completing weekly work and passing the examination should
receive some academic advantage.† Under his guidance the
Syndicate established regulations for a Vice-Chancellor's Certificate. These were awarded to students obtaining six terminal
certificates in any one of three groups:

A – Literature, Language, Political Economy, History, Logic.
B – Political Economy, History, Logic, Moral Science.
C – Natural Science, Logic.‡

In 1883 the University obtained the power to affiliate local
colleges, which meant that their students who qualified for a Vice-
Chancellor's Certificate would be exempt from the Previous
Examination and a year's residence if they entered Cambridge.[73]
In 1886 this privilege was extended to 'any institution founded for
the education of adult students . . . in any place within the United
Kingdom or any part of the British Dominions'.§ This permitted
affiliated local centres the same privilege as local colleges. Derby,
Hull, Newcastle on Tyne, Scarborough and Sunderland were
affiliated in 1887 and Exeter and Plymouth in 1888.[74] From 1907
the Gilchrist Trust assisted affiliated centres financially.‖ Under
these regulations there were now four groups:

* Sheffield Archives Dept., Carpenter Papers 13. The additional figures for
York are the numbers of men – this information is not given for the other two
centres.
† E.g. *Report on 11th Annual Co-operative Congress*, 1879, p. 8 – BEMS 1/16.
‡ *Calendar of Local Lectures*, 1880, p. 148 quoting revised regulations of
1878. See *Annual Report* for 1877 (BEMS 22/1, f. 69). Prof. Kelly (*Adult
Education*, 1970, p. 236) is wrong in stating that the Vice-Chancellor's Certificate was not established until 1895. See also Local Examinations Syndicate,
minute book 3, 15 March 1879 and 25 October 1882.
§ University Archives, C.U.R. 57.2, 142. The two phrases have been trans-
posed for clarity.
‖ Liverpool Univ. Archives, Southport Univ. Ext. Soc. meetings minute
book 1896–1943, f. 18.

A – Literature
B – Modern History
C – Political Science
D – Natural Science.[75]

Students had to follow particular combinations of subjects, which required considerable planning. Some courses were not well supported, and students beginning late or moving to another town sometimes required courses to be repeated. Much of the Local Lecture Committee's time was taken up with such problems.* The number of affiliated centres varied from time to time, but there were still three – Exeter, Leicester and Southport – when the system was abolished in May 1928.[76]

The older pattern of terminal and Vice-Chancellor's certificates was retained for the benefit of non-affiliated centres. In 1894 R. D. Roberts persuaded the Syndicate to change the regulations for these and to institute a sessional certificate for the benefit of pupil teachers.† The sessional certificate could be awarded with honours if the candidate wrote an essay 'involving independent work'. Four sessional certificates and an examination led to the new Vice-Chancellor's Certificate. Sessional certificates were last issued in 1960, but the last Vice-Chancellor's Certificate was issued as long before as 1938.[77] Since many students took the trouble to obtain certificates between 1879 and 1938, it is clear that they served a useful purpose. There is no record of how many affiliated students went to Cambridge to read for a degree. E. T. Clarke of Exeter, who was admitted as a local lecturer in 1905, had gone to Sidney Sussex as an affiliated student.‡ Other students no doubt gained local advantages from their possession of an attractively-printed certificate signed by the Secretary or the Vice-Chancellor.§ However the various efforts made to change the system suggest that it was not entirely satisfactory.

Most of these unsuccessful reforms were proposals from the Syndicate to the Senate for a qualification which was not to be

* Some of the difficulties are given in the *Report* of the 1890 Conference at Cambridge, pp. 6–10 – BEMS 28/2.
† *22nd Annual Report*, 1885, pp. 2, 3, 11. For pupil teachers see above p. 97.
‡ BEMS 36/7, p. 11. In 1894 it was said that five affiliated students had gone to Cambridge (*University Extension Congress Report*, 1894, p. 34). In 1937 a Derby student was awarded a bursary by the Board after passing the examination for a course (Derby Univ. Ext. Soc., *Annual Report* 1936/7).
§ For many years the certificates were printed on parchment by the University Press.

called a certificate. This was rejected by the Senate on the grounds that it gave a privilege which could only be obtained by residence in Cambridge during term time.* As early as 1875 Moore Ede had suggested a degree for extension students to be called A.C. (Associate of Cambridge).[78] About 1896 R. D. Roberts began a prolonged campaign for a Diploma in Arts for extension students. He collected the views of schoolteachers and local lecturers, and in 1897 got a sub-syndicate formed to consider the problem. A group of lecturers which included A. J. Grant and Somerset Cranage put forward an alternative plan for granting the 'title' of A.U.C. (Associate of the University of Cambridge).[79] This was prompted in part by the Government's proposal to transfer the powers of school boards to local councils. Roberts, however, pressed on with his Diploma scheme, producing a *Suggested Draft Regulations for a Diploma for Non-Resident Students.*[80] The sub-syndicate reported in its favour. There were to be three series of examinations to be taken either through Local Lectures or the Higher Local Examinations. Roberts issued a leaflet and flysheets were produced supporting it, but there was strong opposition in the Senate and elsewhere by those who thought it might be mistaken for a Cambridge degree. In May 1898 the Senate rejected the grace for the diploma.[81] No later suggestions for an academic distinction got as far as the Senate. Roberts addressing the Cambridge Summer Meeting in 1908 tentatively suggested an 'associateship', but there was no response.[82] After he became Secretary Dr Cranage wrote an article for the *Cambridge Review* proposing a bolder plan, and as late as 1924 the Local Centres Union was discussing the matter.†

* Even now the Board of Extra-Mural Studies can only issue a certificate.
† 'An External Degree', *Cambridge Review*, 8 November 1918. Local Centres Union minute book, 1912–36, f. 32.

# 9

# The Board, 1924-39

The effect of both World Wars was to reveal deficiencies in the educational system and to stimulate attempts to remedy them. Conscription (and in World War 2 evacuation) showed how many people had failed to benefit from what the State provided, and led to the passing of Education Acts in 1918 and 1944. When the former Act was only a Bill the Syndicate called a conference at Cambridge to consider what might be done for adult education, and in due course a deputation which included representatives of local centres and tutorial classes waited on the President of the Board of Education, H. L. Fisher.[1] This led to the first comprehensive official enquiry into adult education, which began with a sub-committee of a Reconstruction Committee appointed by the Cabinet in July 1917.[2] The main committee soon became the Ministry of Reconstruction under Lord Addison, and the sub-committee became a committee. Albert Mansbridge was a member and Oxford, the W.E.A. and the Labour Movement were much better represented than Cambridge and the Extension Movement.

The first report of the Committee, in March 1918, was concerned with the need for adult education for everyone and the social conditions which hindered this. The second, in July 1918, recommended the establishment of an Army Educational Corps; and the third, in May 1919, was concerned with libraries and museums. It was the fourth and final report which dealt with extra-mural education. This contains as introduction a concise history of adult education from 1800 onwards.* The Committee felt that extra-mural work was still in the hands of amateurs and suffered from the 'discontinuity of much of the work done'.[3] It

---

[1] For References see p. 205.

* This was the standard history until Prof. Kelly's book appeared in 1962. It makes little mention of the Syndicate's work, and gives statistics only for tutorial classes.

suggested a wider range of subjects including natural science, modern languages and (with a vision of Kingsley Amis' 'Merrie England') Arts and Crafts to be taught by the universities.[4] Finally it expected the universities themselves to finance the improvements which it recommended. These recommendations appear to have little regard for the practical aspects of extra-mural work. Attempts to teach sciences and languages had never been particularly successful. The former required laboratories and technical equipment, while the latter required a more intensive training than either tutorial classes or extension courses could provide.* The vision of many returned ex-servicemen clamouring for the study of design 'whether in illumination of manuscript or in needlecraft of many kinds, or in metal work, slate work, etc.'[5] hardly needs further comment. Finally it ought to have been clear that the universities were not able to provide further finance for either tutorial classes or extension courses. The considerable amount which the W.E.A. had already received for tutorial classes had almost exhausted the reserves. Later in the *Report* this point was recognised and extra grants from both central and local government were recommended.[6] Since the Board of Education was revising its regulations to provide improved grants, the Committee was able to exert considerable influence in solving the problem of finance which had troubled the extension movement for so long.

Perhaps the most important recommendation of the Committee, however, was the establishment of separate Departments of Extra-Mural Adult Education, which was adopted by almost every university and university college. The *Report* justified this by the complaints that at Oxford and London the Tutorial Classes Committee was subordinate to the Extension Delegacy or Board. It did not mention the Cambridge alternative by which the two were of equal status and responsible to the Syndicate.[7] The creation of a new department was intended to give independence and more finance to tutorial classes at the expense of extension courses. Unfortunately the creation of separate departments in Redbrick universities has apparently restricted rather than encouraged the work of the W.E.A.[8]

Another recommendation of the *Report* which was adopted by

---

* See *Natural Science in Adult Education* (1927) and the Cambridge plan for language teaching in 1893 (BEMS 38/9).

most universities over the next thirty years was the appointment of resident tutors in areas remote from the universities.[9] Once again the result has been different from the intention. The Committee thought that local colleges might develop as a result, but because most of the resident tutors were appointed for rural areas this has not happened. Finally the Committee suggested that adult students should spend varying periods of time at universities.[10] This was an experiment which the Syndicate had already tried. Neither the Committee nor the Syndicate anticipated the final result – a full-time degree course for mature students.

Although the Ministry of Reconstruction had disappeared in January 1919 before the *Final Report* appeared, the Committee was reconstituted as a committee of the Board of Education. It was a larger and more representative body in this form. Dr Cranage was a member from the beginning.* It issued a series of eleven very useful *Papers* between 1922 and 1933. These included *Recruitment, Training and Remuneration of Tutors* (no. 2, 1922), *The Development of Adult Education in Rural Areas* (no. 3, 1922), *Full-Time Studies* (no. 7, 1927), and *Adult Education and the Local Education Authority* (no. 11, 1933). Despite the difficulties of the years between the Wars, their influence on the development of adult education was considerable.

Soon after the *Final Report* appeared the Cambridge Syndicate considered its recommendations. The Committee wanted 'a department of extra-mural adult education with an academic head' in each university which was:

(i) To promote the further development of such kinds of extra-mural adult education as can properly be assisted by the universities.

(ii) To represent the needs and desires of adult students to the university authorities.

(iii) To report upon questions arising from the work of the university in the sphere of adult education, such as the requirements of new types of student, the value of more educational experiments, and the possibility of extending the influence of the universities into fields as yet untouched by them.[11]

The Syndicate's report to the Senate on 9 June 1920 accepted this.[12] It pointed out that the examining and lecturing work had

* *Adult Education Committee Paper No. 1* (1922), pp. 3, 4. *University Reporter,* 1923–4, p. 123. Dr Cranage still felt that the extension movement was under represented on it (Local Centres Union, E. Anglian minute book, 1914–40, f. 15).

increased so that there were valid arguments for separating the two aspects of the work once more. This was, as we have seen, technically easy, because the two sides of the Syndicate's work had been run separately since 1891. Financially it was more difficult because Local Examinations had subsidised Local Lectures since 1876. The Report stated that

Hitherto the Secretarial and Office expenses of the Local Lectures and of the Tutorial Classes have been mainly charged to the Local Examinations account ... It is becoming increasingly difficult for that account to meet the charge; and the development of the work will make it quite impossible ... all [other] expenses ... have been borne by the Local Centres, and have been met very largely by voluntary subscriptions.

If proper staff were to be recruited and adequate salaries guaranteed to full-time lecturers then both the University and the Government would have to assist. The Syndicate proposed to await the decision of the Royal Commission then sitting. A further complication was the cramped condition of the Syndicate Building. For example, the Local Lectures library had to be kept in the clerks' office on the first floor, on the landing and in the examination clerks' office on the ground floor.* There had also been difficulties with flooding in the cellar where the book boxes were packed.[13]

In 1922 the Royal Commission reported. It recommended a Treasury grant of £6000 a year to Cambridge for extra-mural work, but the full amount was still not being paid two years later. Nevertheless the Syndicate proposed that the division should be made with a separate Board of Extra-Mural Studies:

(a) to organise and superintend the extra-mural teaching work of the University;

(b) to assist in providing financial support for adult students entering the University after a course of extra-mural study;

(c) to co-operate with the Board of Education, Local Education Authorities, Voluntary Organisations, and the Extra-Mural Authorities of other Universities.[14]

There was no opposition in the Senate and the new Board took over in the autumn of 1924. The new body consisted of the Vice-Chancellor and 20 members. Half of them were chosen by the Senate from its own members, a quarter were nominated by the W.E.A., and a quarter by the Local Centres Union. The first Board included Miss Helen Colman (James Stuart's sister-in-law),

* Information from Mr B. W. C. Green, former chief clerk.

Albert Mansbridge, and two former local lecturers, Dr Holland Rose and Mr F. R. Salter. Dr Cranage became the Board's first secretary and the two clerks, Mr Richford and Mr Green, his first staff.[15] Almost the first duty of the Board was to replace the assistant secretary, G. H. Leigh-Mallory, who had disappeared while climbing Mount Everest.[16] From a short list of 50, six candidates were interviewed and Geoffrey Fletcher Hickson of Clare, then an assistant master at Highgate School, was chosen.[17] This decision was more important than anyone may have realised at the time. He was to remain with the Board for the rest of his career and to be its Secretary for forty years.

The Board then turned its attention to the provision of its own offices. The Local Examinations Syndicate had promised to pay for its construction, and Dr Cranage found the task particularly congenial. Before the Board had been formed he had carefully considered sites at the south end of the Examination Schools and at St Andrews Hill near the Masonic Hall – both of which already belonged to the University. He favoured a site in Mill Lane immediately to the south of Syndicate Building. This was part of the garden of Kenmare House in Trumpington Street – an orchard and a croquet lawn – which belonged to the Syndicate.[18] This would have been rather cramped if the University had not been able to use funds provided by the Syndicate to buy the property to the south and east. This was Sindall's builder's yard, most of which is now covered by the Mill Lane Lecture Rooms.[19]

The Board chose as architect George Hubbard of London, a friend of Cranage who had already worked for the University.[20] He was given a specification for: 'Board Room, Library, Secretary's Room, Assistant Secretary's Room, Clerk's Room, Lavatories and Storage accommodation'.[21]

By November 1924 he had submitted plans for the present Georgian-style building, and work started on it in May 1925.[22] Although there was a building committee most of the decisions were taken by the architect and the secretary in collaboration, which resulted in a strong and durable building with the best possible materials. The Syndicate had originally offered £14,000 towards the cost, but had raised this to an absolute maximum of £16,000 in 1925.[23] Numerous calculations by Dr Cranage show his anxiety at the mounting cost and various economies were introduced at a late stage. The roof was not tiled as intended, some of the stone pilasters were omitted and the Board Room was not given an oak floor. Since the building was to be called Stuart House, it was found possible to pay for the Board Room panelling out of the James Stuart Fund and the Misses Colman paid for some of the fittings in his memory. Despite these economies there were complaints from the Building and Finance Committees about the high standard of decoration. J. M. Keynes, when inspecting the new building, remarked that it was mural rather than extra-mural.[24] After a stormy interview with the Finance Board on the subject Dr Cranage aptly remarked that in twenty years time the University would be grateful to him for building so well.*

The building was mainly intended as offices although the library was to be 'the special home of the extra-mural students resident in the University'.[25] This led to Albert Mansbridge writing a curious article in *The Times* to celebrate the opening of Stuart House on 5 February 1927. It began:

POOR STUDENTS
A NEW CAMBRIDGE FOUNDATION

In loyalty to post-Reformation tradition, Cambridge in yet another matter of importance, the foundation of a House for Extra-Mural Students, has started ahead of her elder sister Oxford, for whom she offers the anxious prayer 'God speed you behind us' . . .[26]

The article appears to describe the foundation of a second establishment for non-collegiate students rather than administrative offices of the Board. There is no record of Somerset Cranage's comments on this description of his achievement.

In the following year Dr Cranage returned from a tour of Canada and the United States to find amongst a pile of circulars which his housekeeper had put on one side a letter from Stanley Baldwin offering him the deanery of Norwich.[27] This he accepted and resigned the Secretaryship in March 1928. As an expert on medieval architecture the new appointment pleased him, and he was able to rebuild the Lady Chapel of the Cathedral as a war memorial and to clean and repair the cloisters. He also became chairman of the Central Council for the Care of Churches and the Cathedrals Advisory Committee.† Although the Board had some

---

* This was proved in World War 2 when almost no maintenance was required. The building is only now becoming too small for the Board's administration.
† *Annual Report of King's College Council*, 1957–8, p. 26. He resigned the deanery in 1945 and died at Windsor in 1957.

hesitation about his youthfulness, Geoffrey Hickson succeeded him as Secretary of the Board. He was only 28 at the time, the son of Professor Hickson, F.R.S. He had taken a history degree at Cambridge and won the Gladstone Prize in 1924. As well as being Secretary of the Board, he was also a member of the Council of the Senate for 16 years and a University member of the City Council since 1943. In 1947 and 1962 he was Mayor of Cambridge and has been an alderman since 1952. In 1963 the new Fitzwilliam College elected him a Fellow.

The division between Local Examinations and Local Lectures (both administratively and physically), followed by the retirement of Dr Cranage marks the end of an era in university extension work at Cambridge. For over half a century the administration had been in the hands of no more than four people – a secretary, an assistant secretary, a clerk and an office boy.* It had been a very personal affair and the relationship with local centres was a very close and friendly one.† Under the new Secretary the work grew larger and inevitably more impersonal. The variety of work undertaken by the Board increased while the number of universities and voluntary bodies concerned with extra-mural work grew. No longer could national arrangements be settled quickly by a meeting between the Cambridge Syndicate and the Oxford Delegacy. Local education authorities now had to be consulted more frequently and persuaded to co-operate with the Board. It became impossible for one man to control everything so closely. During the next twenty years a two-part solution was evolved. As more money became available extra staff were appointed to organise as well as lecture. Meanwhile the Board, freed from the Syndicate's concerns with Local Examinations, were able to take a more active part in the work. This development was greatly assisted by the Board's decision when appointing Geoffrey Hickson to give him the support of a strong chairman.[28] But it was his own skill and charm which enabled the old friendly atmosphere to be retained under the new arrangements and which prevented the problems of expansion becoming crises.

As we have already noted after 1919 the number of authorities

* Mr B. W. C. Green described to me how at the end of the Summer Meeting he was sometimes left in sole charge of the office.

† This can be seen in the references in University Extension Society minutes to meetings with 'our beloved Syndicate Secretary'.

and organisations concerned with adult education increased rapidly. Professor Kelly has shown that Oxford, Cambridge and London provided almost all the extension courses in that year, but by 1938 they provided only two-fifths.[29] In 1924 the Board had local centres as far away as Plymouth, Leicester, Southport and Sunderland. By 1939 it had only five centres outside East Anglia – Barnstaple, Derby, Hastings, Rugby and Southport. By 1973 the last of these, Derby, will have been given up. Cambridge has always recognised the right of each university to its own 'territory' and the inevitability of withdrawal, but it has usually required considerable tact and careful negotiation before each surrender. Most local centres have deplored the change. In 1942 a Lincolnshire centre secretary threatened to resign if courses were to be run by Nottingham; in 1954 Southport regretted their transfer to Liverpool; in 1962 the Northamptonshire centres petitioned the Board against transfer to Leicester, and the Wellingborough Extension Society eventually preferred to dissolve itself.[30] The other universities took a strong line about these relics of past arrangements. They had appointed a resident tutor for the area and the largest town was the responsibility of another university. In 1942 when the Board suggested that it should collaborate rather than compete with Exeter University College in North Devon, the latter rejected collaboration 'in any sense' and asked Cambridge to leave its 'territory'.[31] Although the historic reasons which brought Cambridge to these towns were no longer valid, the centres did have a reasonable grievance in some cases. The lecturers provided by Redbrick universities were usually paid considerably less than Cambridge lecturers, therefore a transferred centre was debarred from hearing Cambridge lecturers for the future unless special arrangements could be made.[32]

As opportunities for the work diminished elsewhere, the Board turned to providing courses in its own 'territory' of East Anglia and the East Midlands. Originally little had been done in this area. Bedford was the first local centre established in 1875, but it was soon given up to Oxford. Norwich began in 1877 and it was followed by Cambridge in 1887 and Ipswich and Colchester in 1889. Even in 1924 the Board had only 20 out of 42 extension courses in what was later to be its territory.[33] Tutorial classes were not organised over the whole of England, but many were in places distant from Cambridge. This apparent neglect was principally due to a lack of demand and finance. East Anglia contains no large industrial towns of the kind where Stuart began extension work, and it had few large seaside resorts whose inhabitants could pay for courses. It was not until subsidies became available that the Board was able to stimulate demand through its Rural Areas Committee.

In 1887, 1890 and more recently the Syndicate had experimented with work in rural areas,[34] but it had never been particularly successful. The cost of providing a course at Dry Drayton is little different to the cost at Derby, but the total audience available is much smaller. One means of overcoming the financial difficulty was a continuation of the 'Norfolk Experiment' whereby teachers were trained at the cost of the County Council to lecture in rural areas.* In 1920 the Norfolk County Council were again paying the Syndicate to provide special courses, and in 1921 East Suffolk followed their example.[35] It was about the same time that Professor Peers developed at Nottingham the plan of resident tutors partly financed by local authorities which was generally adopted.[36] The W.E.A. began similar work in Bedfordshire and the Board in Cambridgeshire in 1927. In Cambridgeshire the Board took over from the Rural Community Council the scheme of short courses which it had already organised, but was unable to continue. The Cassel Trust made a three year grant to the Board to provide single lectures and short courses in rural areas.[37] When the grant ended in 1930 the W.E.A. also wished to give up their work in Bedfordshire from lack of money. The Board then decided to enlist the support of local authorities in carrying on in both counties and eventually extending to other counties. The Rural Areas Committee was appointed with provision for local authorities to be represented. Bedfordshire were the first to offer help to appoint a resident tutor and in 1931 Cambridgeshire offered £100 a year for a resident tutor.† Because of cuts in Government grants to local authorities at this time no further counties agreed to join for several years. However the Board was able to continue with grants from various educational charities.[38] In 1934 the Isle of Ely and Essex County Councils began to contribute to the work and in

* See p. 90 above. The original experiment seems to have been abandoned about 1907, although in 1909 the Council was still paying for teachers to attend Norwich courses (Norwich Record Office, Univ. Ext. Soc. meeting minute book, 17 November 1910). It was revived after World War 1.

† BEMS 10/1. The Bedfordshire tutor was Harold Shearman, afterwards Director of Education for London County Council and knighted.

1938 it became possible to appoint resident tutors in Norfolk and Essex. Hertfordshire provided funds for a W.E.A. organiser in rural areas in 1941, but it was not until the end of World War 2 that the scheme could be extended to Northamptonshire and Suffolk.[39]

From single lectures and short courses the work progressed to terminal and sessional courses and eventually to tutorial classes. In addition from 1932 to 1939 the Board organised a Saturday School in Cambridge each year for students from rural areas. The first of these on 18 June 1932 was an experiment for 'members of classes in Bedfordshire and Cambridgeshire' – the only two counties where rural area work was being carried on. More than 250 students applied. In the morning the Secretary talked to them about Cambridge and they visited some of the colleges. In the afternoon they heard a lecture on 'An Economist looks at the World'.[40] No charge was made, but the students had to pay their own travelling expenses and buy their own lunches. In the following year the School was held in May and a charge was made for tea. Because many students were not free until noon the two lectures – 'New Germany and Europe' and 'A Journey into Space' – were held in the afternoon and evening.[41] This pattern was followed until the outbreak of War and proved very popular.

This work in rural areas was only made possible by the subsidies which had become available. It had been possible since the Education Act of 1902 for local authorities to contribute towards the work of the Syndicate, but for many years few did so. In February 1915 the Syndicate had a leaflet printed for circulation to all centres showing which local authorities had helped.[42] Two ran the local centre and met any deficit, eight gave a grant, six a guarantee, and six helped in other ways, usually by providing a room. In addition it was possible to apply direct to the Board of Education for a grant and three centres did so in 1915. This meant that 23 centres out of about 35 received some assistance.* In 1921 the Syndicate estimated that extension courses received subsidies of £37 from the Board of Education and £923 from local authorities out of a total expenditure of £5149.[43] They felt that local education authorities were becoming 'increasingly sympathetic'.†

An Education Act in 1921 allowed the Board of Education to draw up Adult Education Regulations on a new basis.[44] Chapter II

---

* Two centres, Leicester and Portsmouth, received local and central grants.
† Local Centres Union, minute book 1912–36, f. 22.

of these regulations was concerned with university extra-mural courses which were defined as:

(i) Classes Preparatory to Three Year Tutorial Classes;
(ii) Three Year Tutorial Classes;
(iii) Advanced Tutorial Classes;
(iv) Tutorial Class Vacation Courses;
(v) University Extension Courses.

The numbers in the classes were restricted to 24 or 32 and so long as certain minimum requirements were met the Board would make a grant of up to £60 a year. Extension courses were to have between 12 and 32 students for a maximum grant of £45. The regulations were drawn up with W.E.A. students principally in mind and the difficulties of obtaining an extension grant on these terms deterred many societies from doing so.* Chapter III of the regulations was for courses given by 'approved associations':

(i) Terminal Courses;
(ii) One Year Courses;
(iii) Vacation Courses.†

The regulations also allowed a local authority to take over the entire responsibility for a course.

Complaints from the Extension Movement eventually produced revised regulations in 1938.[45] Chapter III courses were now defined as:

(i) Three Year Tutorial Classes;
(ii) Advanced Tutorial Classes;
(iii) Tutorial Class Vacation Courses;
(iv) University Sessional Classes;
(v) University Extension Lecture Classes;
(vi) University Extension Lectures.

The tutorial class regulations were hardly amended, but there were considerable changes for extension courses. Separate grants of 37s 6d could be obtained for both lecture and class. The former required a minimum of 75 students and the latter a maximum of 32. Although they were a great improvement on the 1924

---

* If the numbers exceeded 32 as they usually did, then a special register of 32 selected students had to be kept. See Local Centres Union minute book 1914–40 for the frequent complaints.

† Chapter III courses (as they were known) survived until 1946. They were sometimes organised by the universities on behalf of the W.E.A. and other voluntary bodies.

Regulations these were still restrictive and many extension societies refused to apply for a grant until financial reasons obliged them to do so.

The differences in the regulations between extension courses and tutorial classes was reflected in the division between the work of the Lectures Committee and the Tutorial Class Committee of the Board. In practice there was little real difference away from Cambridge. Local lecturers could also act as tutors, even in the same towns.* Extension courses could succeed tutorial classes and *vice versa*,† but still each committee had its own list of teachers and co-ordination was apparently achieved only through the Secretary's efforts.‡ The Tutorial Classes Committee organised its separate Summer School each year, and its tutors met in conference separate from the lecturers. For a time there was even a separate Tutors' Association.[46] In 1932 the Cambridge Local Lecturers' Union approached the tutors to form one organisation and even changed its name to the Extra-Mural Lecturers' Union without any success.[47]

In accordance with W.E.A. practice the Eastern District secretary was established at Cambridge in order to maintain contact with the Board. The first secretary was G. H. Pateman, who also served as joint secretary of the Tutorial Classes Committee from 1920.§ In February 1922 the W.E.A. District Committee approached the Tutorial Classes Committee for an honorarium for Mr Pateman.[48] No action was apparently taken at this point because he was spending two years at Trinity as James Stuart Exhibitioner. By May 1923 there were serious doubts about whether the District could continue to employ him and on the establishment of the Board he was given an annual honorarium of £150. When Stuart House was opened he was given an office there and allowed to use it for W.E.A. work.‖

* E.g. Mr Hardman took a preparatory class at Wellingborough in 1925/6, a tutorial class from 1928 to 1930, and an extension course in 1934 (leaflets in Mr Hardman's possession).

† E.g. Halesworth in 1931 and Letchworth in 1932 (*Cambridge Bulletin*, no. 8, pp. 2, 15).

‡ As early as 1920 there had been a plea for co-operation (Parry, ed., *Cambridge Essays on Adult Education*, pp. 167, 168). From 1934 to 1946 both Committees organised courses in Wellingborough simultaneously although the population was only 30 000.

§ Previously Albert Mansbridge had been joint secretary.

‖ BEMS 6/1, 173, 211. BEMS 4/1, 3. BEMS 7/1, 24. In 1927 the honorarium was increased to £200 (BEMS 6/1, 223).

The next financial crisis in the Eastern District arose in 1935 when Mr Pateman found his two posts burdensome and proposed to resign his joint secretaryship. Since the District was unable to replace his honorarium, he was persuaded instead to become assistant secretary to the Board together with Mr Hardman.* The District then appointed Mr F. M. Jacques as its secretary and, through the Tutorial Classes Committee, asked for assistance in paying his salary. The Board now had a Finance and General Purposes Committee which expressed some doubts about the legality and expediency of such subsidies. However it recognised the special circumstances of Mr Pateman's transfer and agreed to a single grant of £100 to cover this.[49] The Board also continued to provide office space until a District Office was established after World War 2.

As soon as this problem had been solved difficulties arose about the respective roles of the District and the Board. The Board, like the Syndicate, had always taken direct financial responsibility for tutorial classes. It appointed and paid the tutors and in return received the Board of Education grant. If the class failed to obtain a full grant (which happened not infrequently) the loss was the Board's. The W.E.A. paid a nominal sum to the Board for each class,† and the small local expenses. All administrative costs were met by the Board. The trouble began at Ipswich where there was an active W.E.A. branch. A class in psychology which began there in 1935 had too many students.‡ A second class in the subject started in 1937 could not have the same tutor and the branch lodged a formal complaint:

that the problem of attracting students into W.E.A. Classes is being made increasingly difficult by the failure, on the part of the Extra-Mural Board authorities at Cambridge, to recognise the principles on which the success of Adult Voluntary Evening Education depends . . .[50]

The Tutorial Classes Committee (on which the W.E.A. was well represented) replied emphasising their responsibilities for the

* BEMS 7/1, 108. Information provided by Mr Hickson. The post of assistant secretary had not been filled in 1928.

† In 1936 this was £1 for the Ipswich class, £1 10s for the Rugby class and nothing for the Cambridge class, which met at Stuart House (BEMS 6/2, 30).

‡ BEMS 6/2, 14. Psychology (and psycho-analysis) was a very popular subject with tutorial classes at this time. The W.E.A. disapproved of the introduction of psycho-analysis to a class.

provision of tutors in a letter signed by Mr Pateman, Mr Jacques and Mr Hickson.*

In the following year there was a disagreement between members of the Board about the work in rural areas. Because the W.E.A. had helped to start this work in Bedfordshire, it had always been represented on the Committee. Now Mr Ernest Green suggested that all the responsibility should be given to the District and the Rural Areas Committee should merely act as co-ordinator. Financial responsibility was not mentioned but would presumably have remained with the Board. The District complained that the existing arrangements weakened the W.E.A. because the Board undervalued its work and had taken over the provision of 'the more elementary courses'.† Unfortunately the District had neither the staff nor the finance to continue the work and the˙chief reason for its complaint seems to have been an attack launched on the District by the National Council of Labour Colleges.‡ The W.E.A. memorandum to the Board said:

Active hostility to the W.E.A. has this year been met and overcome at Chelmsford, Lowestoft, Yarmouth, Hoddesdon, etc. The attacks met there and elsewhere result from an intensified campaign on the part of the N.C.L.C. among trade unionists, that body setting out to persuade them that the state grants for classes received by the W.E.A. and our co-operation with the Universities (which latter bodies do not want an educated working class movement!!) are reasons for mistrust and avoidance of the classes we organise.[51]

The Board appointed a sub-committee to consider its relationship with the W.E.A., and the agreement reached was a satisfactory compromise. The District received increased representation on the Board and there was increased co-operation and co-ordination between the officers.[52] In 1940 there was a complaint from the W.E.A. against the secretary of the Tutorial Classes Committee which was disposed of by another sub-committee.[53] In February 1941 the Ipswich branch made its final appearance in the minutes by refusing to hand over part of its grant from the borough council for a class which the Board had provided.[54] During all the distrac-

* BEMS 6/2, 86. BEMS 48/10. It is possible that the class objected because it wished to include psycho-analysis in the course (see BEMS 6/2, 90 and 106).
† I.e. chapter III courses. BEMS 48/11 – W.E.A. and Mr Hickson's memoranda.
‡ For this body and its work with trade unions see Kelly, *Adult Education*, pp. 283, 380.

tions of World War 2 and in the changed post-War conditions, most of the old disagreements and the old differences were forgotten.

During the same period the Board also felt the effects of the economic depression in many ways. Lack of funds hampered some of the work, but it was also able to do something to assist the unemployed. A victim of the times was the Board's *Bulletin of Extra-Mural Studies*. This was a distant descendant of the *University Extension Journal* established by R. D. Roberts in 1890. Until 1895 this had a rival in the Oxford *Extension Gazette*, but in that year Oxford, Cambridge, London and Victoria came together in a new series of the *University Extension Journal*.* This printed the programmes of all four universities together with notices of general interest, short articles,† book reviews and reports from local centres. There were difficulties with finance (it sold at three-pence a copy) and with the amount of local information. From 1904 to 1907 it was called *University Extension* and from 1907 *University Extension Bulletin*. Victoria withdrew from the group and the number of issues was reduced to three a year instead of nine. It appeared in three different editions containing local notes for one university only.

After World War 1 the practice of separate editions was abandoned and the *Bulletin* was only published twice a year at the beginning of the Lent and Michaelmas terms. In 1925 Dr Cranage suggested that changes were needed to meet the proliferation of authorities. The British Institute of Adult Education also had plans for its own journal. His suggestion that the *Bulletin* should be published by the Institute or by all the extra-mural departments was rejected and the Board decided to produce its own *Bulletin* from Michaelmas 1926.‡ This was a larger publication than its predecessors and contained separate sections on extension work and tutorial classes (with different editors) and a list of all the Board's courses and classes.§ Like its predecessors the *Bulletin* is full of useful information. Unfortunately it was never self-

---

* Complete sets of all these publications, except the Oxford *Gazette*, are preserved in the Board's archives (BEMS 42/1–22).

† E.g. Mansbridge on 'Co-operative Education' in 1899 (vol. 4, p. 138).

‡ The Institute began publication of its *Journal* about the same time.

§ Although it still contained 24 pages, they were all now devoted to Cambridge news. It was also better produced – by the University Press. See BEMS 48/1, and 34/18, 18 March 1926.

supporting. In Michaelmas 1928 the price rose threepence to six-pence and it was decided to issue one number a year. Its publication was reluctantly suspended by the Board after no. 9 had appeared in November 1932. Although not intended to be a permanent suspension, it never re-appeared.

Little has been written about the work of the Extension Movement for the unemployed between the Wars.* Although the Board did not organise special courses until 1936, it was active in the field much earlier through its local centres. It helped to found one of the educational settlements which did so much work in this area. This was the Percival Guildhouse at Rugby which was founded in 1925 as a memorial to a former headmaster of Rugby and 'to stimulate and satisfy the demand for Adult Education in Rugby and District'.† Rugby was a Cambridge centre for tutorial classes and extension courses, and the Board not only nominated a representative on the Guildhouse Council, but gave £200 a year for the Warden's salary.‡

As early as 1929 the Norwich University Extension Society had helped to organise 'Instructional Classes for the Unemployed', and in 1932 the Southport Society made arrangements to admit the unemployed to its regular courses.§ Other centres no doubt made similar arrangements where possible. Then in 1936 the Board was approached to provide courses for residential centres for the unemployed in Cheshire and Durham. From then until 1939 lectures were given at Wincham and Hardwick Halls. The Board paid the lecturers' fees and part of the travelling expenses; the centres only provided hospitality.[55]

It was not only the difficult economic conditions which kept the number of Cambridge courses and classes low between the Wars. The steady reduction of centres outside East Anglia and the East Midlands also helped, and there were rival attractions in the cinema and the radio. In 1927 the secretary of the Middlesbrough centre wrote that: 'there never was a time when it was more difficult to keep the Centre going, due chiefly to a large portion of

---

* Prof. Kelly does not discuss the topic in his *History of Adult Education*.

† BEMS 38/21. *Cambridge Bulletin*, no. 1, p. 4. Dr Percival when Bishop of Hereford had ordained Somerset Cranage.

‡ BEMS 4/2, 26. The Board's assistance is not noted in Kelly, *Adult Education*, p. 278.

§ *Cambridge Bulletin*, no. 6, p. 5. Local Centres Union minute book 1912–36, ff. 44 and 48. Liverpool Univ. Archives, Southport Univ. Ext. Soc. meetings minute book 1896–1943, f. 61.

the population living out of town, and another portion finding the " Pictures " more to their liking '.[56]

Some centres even advertised their courses in the local cinema,[57] and there was even closer co-operation with the radio. As early as March 1926 the British Broadcasting Company had suggested that the Board might ask for a local station for Cambridge.[58] When the British Broadcasting Corporation was formed in 1927 it established an Adult Education Section under a former W.E.A. tutor and the Board was duly notified.* The B.B.C. also appointed a committee of enquiry into the subject. Its report published in the following year[59] suggested group listening to talks followed by discussions. Three Cambridge tutorial classes had already arranged an experiment on these lines when their tutor, Mrs M. G. Adams, had given six radio talks on 'Problems of Heredity'.[60] The Cambridge class met at Stuart House with a 'receiving set' borrowed from a fellow of Kings'. 'Discussion proved easy and led to various demonstrations of method', so that the experiment was considered a success.[61]

Although Dr Cranage himself gave talks on the radio and Sir John Reith, Director-General of the B.B.C., was invited to speak at the 1930 Summer Meeting,[62] the proposed co-operation was not very successful. No doubt Mrs Adams succeeded because she was already known to members of the classes and afterwards visited them again, but other discussion groups did not have this advantage. The Local Centres Union complained that talks were usually broadcast in the afternoons only.[63] In 1935 after the amount of educational broadcasting had been considerably reduced the Board was again asked to co-operate with the B.B.C., but even less seems to have been achieved.[64]

As soon as the Royal Commission had expressed its support for the Adult Education Committee's recommendations on full-time studies for adult education students,[65] the Syndicate produced an enlarged scheme for the purpose. The money was obtained mainly from the colleges, either as gifts or as scholarships reserved for adult students on the recommendation of the Syndicate. An early attempt to get financial help from the trade unions was not particularly successful.† Although everyone concerned with the

* *Cambridge Bulletin*, no. 3, p. 71. Information from B.B.C. Written Archives Centre.

† BEMS 8/1, 2, 5, 9, 31, 32. A detailed account of the cost for an adult student can be found in *Cambridge Bulletin*, no. 2, p. 36. Cambridge gave a grant of £225 a year – higher than other universities in 1925 (*Full-Time Studies*,

scheme felt that a short stay of one or two years at the University was sufficient, it soon became clear that most of the eligible students wished to stay for three years and obtain a degree.* This problem had already arisen in the limited scheme for tutorial class students, who had been urged to return to their previous occupation, however menial.[66] The Board probably persisted in this attitude for longer than now appears reasonable because of the difficulties which bursary students faced in obtaining new employment. A pre-World War 2 survey showed that many of them faced a difficult period of unemployment after leaving Cambridge and had to be assisted by the Board.[67] However the students' view was that 'no sane person would re-enter a Coal Pit as an employee if he could avoid it', or 'locals would have regarded [a return to previous employment] as a sign of failure'.[68] Apparently only one student 'never regarded the obtaining of a job as the primary end of education. Though I derived no pecuniary benefit from my stay at Cambridge, my value as a social being has been enhanced thereby.' Many of the first students afterwards regretted that they had not been allowed to take a degree. This gap between the thinking of the University and the adult students shows in itself how important it was to bring the two sides together.

Inevitably the Board was obliged to alter its policy about degrees, and when early in 1928, it learned that the University was considering the position of mature students it asked that the bursary students should be given exemption from the Previous Examination.[69] This removed a considerable obstacle from their path to a degree and was a reversal of the Board's policy of not allowing students to take the Previous Examination without special permission.[70] This was in a way a further extension of the privilege of affiliation devised by James Stuart.† There was no opposition to this proposal in the Senate and it was approved on 15 May 1928.[71] From then onwards every student read for a degree.

Between the Wars the demand for bursaries was very high. In 1931 there were 49 male and 9 female applicants for two bursaries.

1927, p. 30). Unions preferred to give help to individual members of their own union.

* The 1908 report had anticipated this difficulty to some extent by proposing that Oxford should institute special diplomas (*Oxford and Working-Class Education*, 1908, pp. 78–80).

† See p. 138 above. *University Reporter*, 1927–8, p. 895. A copy appears in BEMS 8/1, 40.

They came from all over England and Wales* and ranged from civil servants and managers to coalminers and tram conductors.[72] More applications were received from those attending tutorial classes than extension courses despite efforts to correct the imbalance.[73] Those selected formed a representative sample, except that the proportion of women chosen was somewhat higher.† The average age was below 30 for both applications and acceptances.‡ This was considerably lower than the average age in courses and classes, but applications came only from the younger members and, in general, from those without professional qualifications.

The difficulties which these students encountered were considerable despite the assistance of the Board, their tutors and their colleges. Some had been unemployed for considerable periods before going to Cambridge and were frequently in poor health.§ The Board found it necessary to give every candidate a medical examination. Some had dependents and all needed some vacation work to help their finances. There were difficulties in obtaining additional support from local education authorities – which often involved the Secretary in considerable correspondence – or from charities. At the end of their course there was often no work either in the student's former occupation or in a new field and some had to be assisted by loans from the Board.‖ Yet few of the students regretted the time spent at Cambridge where:

every opportunity exists for the student to pursue his interests. Apart from the libraries one can always turn to lecturers or supervisors for help or advice, whilst the numbers of literary, musical and dramatic societies, debating societies and discussion groups of all kinds is legion.[74]

* It was not necessary to have attended courses arranged by Cambridge to be considered for a bursary (BEMS 8/1, 1).

† Approximately 20 per cent and 25 per cent. At this time the proportion of female undergraduates in Cambridge was only 10 per cent (S. C. Roberts, *Introduction to Cambridge*, Cambridge, 1943, pp. 81, 91). See also BEMS 8/1, 2.

‡ The Adult Education Committee thought 25–35 was the best age for bursary students (*Full-Time Studies*, 1927, p. 21).

§ Two died at the end of their course, and several others caused the Board considerable concern and the expense of treatment (BEMS 56/1–7).

‖ A Joint Consultative Committee was established to find them posts (*Cambridge Bulletin*, no. 8, p. 3).

# 10
# Years of Change, 1939-73

In 1950 when the Board submitted its estimates for the next quinquennium, its submission opened with the statement that:

The demand for adult education is to some extent a reflection of popular interests and of the social and economic circumstances of the country at a given time. It is not easy, therefore, to forecast six years in advance what will be the trends of interest in the Eastern Counties, in which most of the Board's teaching is now concentrated, as they may depend upon factors over which the Board have no control.*

This statement epitomises the work of the Board since 1939. The mere fact that it was necessary to make such estimates indicates the changes which had occurred. The Board was also attempting to plan developments rather than react to changed circumstances. This was inevitable because for the first time most of the finance for its work came from the State. Furthermore it shows that the Board was giving up all its centres outside East Anglia, and would become in one sense just another of the twenty or so departments of adult education in England with its own territory. It also shows that the rate of change was now much greater than either the Board or the Syndicate had ever known.

The increased financial resources of the Board enabled it to appoint more staff, organise more courses and open up new lines of development. The recurring financial crises of the country often spoiled these efforts just when the Board was obtaining good results. During the War and the six subsequent years the work of the Board was hampered not only by shortages of all kinds but also by bad weather and power cuts.† This was followed by a period of financial stringency and the arrival of television as a rival to adult

---

* *University of Cambridge. Needs of the Faculties, Departments and other Institutions* (1952), p. 269.

† *23rd Annual Report*, 1947 contains several stories of the bad winter of the previous year. See also Colchester Univ. Ext. Soc. minute book, 1941–55, f. 33.

education.* More recently, too recently for its effects to be assessed, the Open University has been founded to offer the adult student the prospect of a part-time degree. This is a privilege which University Extension has never been able to offer.

Despite all these problems the Board has expanded its work both in quantity and variety throughout the period. During the War it began to provide lectures for the Armed Forces. This was expected to cease after the War, but it continued on the same scale for some years and is still an important concern of the Board. About the same time it began to provide residential courses, and the demand for these has grown more rapidly than the Board can supply the facilities. There has been a similar increase in the courses for foreign students which the Board began before World War 2. The scheme of bursaries for adult students, however, which was successfully revived after the War has now practically disappeared from lack of demand. Many of the old distinctions and rivalries have disappeared too. Tutorial classes no longer have to be organised through the joint committee with the W.E.A. and extension courses can be organised by that committee. Rural area work as a separate category came to an end in 1945, so that it is no longer necessary to debate whether Sandy (with a population of 5000) is the concern of the Rural Areas or Local Lectures Committee.[1] The newer universities, having established a right to their own territory, are now less suspicious of Cambridge and willing to co-operate with the Board in residential courses.

The initial effect of World War 2 on adult education was somewhat similar to that of World War 1. Some extension societies and centres closed down. War work and evacuation removed some of the local supporters,† while the blackout and fears of enemy raids made students reluctant to attend courses. Even the Munich crisis of September 1938 had a noticeable effect on the Board's work,[2] and in the following autumn, despite the Board's efforts, only nine terminal courses and five short courses were held and only two tutorial classes were begun.[3] However as early as 1938 there had

---

[1] For References see p. 206.

* Bedford University Extension Society had reduced attendances which they attributed to the counter-attraction of television from 1960 onwards (information supplied by the Society).

† The evacuation of schoolchildren from Lowestoft, for example, removed many teachers who were supporters, while the lecturer chosen was fully engaged in war work. Therefore the first wartime course had to be cancelled.

been some indications of the part which the Board was to play during the War. The Wellingborough Society had chosen one of the courses on 'World Affairs' which were popular in Cambridge centres during the Thirties. Because the Munich crisis aroused interest in the subject it was 'considered one of the most successful courses ever held ... It was noticed that a large number of people attended who had hitherto not had any interest in University Extension work. The attendance was larger than it had ever been and for some lectures the audience numbered over a hundred.'[4]

As a result Wellingborough took a further course on the subject in September 1939 with the encouragement of the Board and the Ministry of Information. The Ely centre had a similar experience. It was very reluctant to hold a course in the autumn of 1939 and did not begin until November, but it was later described as a very successful course.[5] By that time the Board was convinced that the effects of the War were not so serious,[6] and, when conditions deteriorated later, courses and classes continued to be held. At Derby, in the winter of 1940–1, an English Literature course was held with a background of 'sirens and gunfire', but attendances were higher than usual.[7]

The official attitude towards adult education was more enlightened in 1939. The Ministry of Information issued a series of cheap illustrated pamphlets on different aspects of the War, and encouraged debates on the issues involved. The Norwich Extension Society organised lectures on behalf of the Ministry,[8] and the subjects of many courses were related to the War. But there was also an increased public awareness of cultural matters at this time. In Michaelmas term 1944 when Bertrand Russell gave a short course at Cambridge on 'Philosophy and Politics' there were long queues outside the Arts School where they were given, and the course had to be repeated.[9] Elsewhere enthusiastic evacuees who had little else to occupy their evenings in rural districts established new Cambridge centres. Music was a popular subject at this time and if it was chosen for a tutorial class then there were usually more applicants than places.[10] For a few years, despite the many difficulties, there was a second flowering of University Extension.

As the War continued the Board also faced increased demands for lectures for the Armed Forces. During World War 1 this duty had been shouldered by the Y.M.C.A., but that body was now able to enlist the aid of a much larger body of adult education organisa-tions.* The first efforts were made for the Militia – the conscripts of the period between Munich and the War – but little had been done before September 1939. Soon afterwards on the initiative of the Y.M.C.A. a Central Advisory Council for Adult Education in H.M. Forces was established, and regional committees to provide the facilities. In February 1940 the Master of Balliol (A. D. Lindsay) wrote to *The Times* to publicise the scheme.[11] In the same month the Vice-Chancellor of Cambridge summoned a meeting of representatives of the Board, the Eastern District of the W.E.A., the Y.M.C.A. and the Army to establish an Eastern Counties Committee.[12] This committee began work on 1 March with its office at Stuart House. At first Geoffrey Hickson acted as the secretary, but later when additional staff had to be recruited Mr Scarlyn Wilson, one of the Board's lecturers, became assistant secretary with special responsibility for the Forces.[13] Because the Forces paid for each lecture or course after it was delivered, there was no money available for immediate expenses and the work was begun with a grant from the Central Council and a loan from the Board.[14] Towards the end of the first year the Haining Report established the right of the troops to an adult education programme and considerable progress was possible.† A large panel of lecturers was built up and lists (similar to those of extension lecturers) were published by the Board.‡ Because this was not strictly one of the Board's activities and because of the difficulties caused by military service,§ single lectures and discussion groups were more frequent than courses or classes. However troops were encouraged by the Board to attend its ordinary courses where this was possible.‖

There were some problems in the early years. The organising tutor for Norfolk reported one incident to the Board in February 1940:

* For work done in World War 2 and before see T. H. Hawkins and L. J. F. Brimble, *Adult Education, the Record of the British Army* (London, 1947).

† See article by G. F. Hickson in *The Cambridge Review* for 13 June 1942.

‡ *Eastern Counties Committee . . . List of Lecturers and Subjects*, Autumn 1941 (BEMS 57/1). Supplements and revised lists were issued later.

§ Often a carefully-arranged course failed because all the prospective students were moved to other areas or countries.

‖ The East Anglian District proposed that troops should be admitted to courses without any charge (Local Centres Union, E. Anglian minute book, 1914–40, f. 60). At Lowestoft sailors attended the Society's courses (information from Miss A. R. Murray and Mr J. Aspinall, 1971).

At the first meeting of my Carbrooke class ten men from the R.A.F. Station at Watton were present; at the second meeting seven including a corporal of the Station Police. At the third meeting, on February 14, only two: Flight-Sergeant Davidson – a man of 19 years R.A.F. service – and the corporal.

After this meeting Flight-Sergeant Davidson was summoned to the Guard Room and taken under arrest before the C.O. of the Station, who forbade him to attend further meetings of the Carbrooke Class. Davidson asked whether this prohibition applied to the W.E.A. class under Mr Scott at Watton; the Group Captain replied that it did ...*

When David Hardman went to lecture to troops at Southend he was first interviewed by the commanding officer and told he was only allowed in because 'the bloody War Office says you can'.† In 1941 the Army objected to two of the Committee's lecturers who were pacifists. The Committee had always refused to employ conscientious objectors who were liable for military service, but both lecturers were Unitarian ministers and neither had attempted to mention their views.[15]

Despite these problems the number of lectures and courses continued to increase. Most of the work was for the Army, and later for Canadian and American troops.‡ The aerodromes which were built in large numbers in East Anglia were also visited, but little could be done for the Royal Navy. There are few large ports in East Anglia and for much of the War only light forces were based in the North Sea. There was a small petrol allowance for lecturers, and to husband this they often went on two or three day trips giving up to seven or eight talks during the day at aerodromes or anti-aircraft sites, and ordinary civilian courses in the evening.§

Later in the War the Committee organised courses in arts and handicrafts and in mothercraft (for members of the W.A.A.F.). Emphasis was placed on residential courses to train troops to give lectures themselves. When the Army Bureau of Current Affairs began to issue its pamphlets, *Current Affairs* and *War News*, courses were organised for officers to explain how these could be

---

* BEMS 57/1. The names of those involved have been omitted from the quotation. The course at Carbrooke was on the Expansion of Germany 1917–39 and that at Watton on the Problems of Peacemaking. See BEMS 4/2, 382.

† Information supplied by Mr Hardman.

‡ For their residential courses at Cambridge a special paperback edition of S. C. Roberts' *Introduction to Cambridge* (Cambridge, 1934) was produced. See also BEMS 57/4.

§ Information supplied by Mr Hardman.

used for discussion groups.[16] By January 1944 the Committee employed five full-time teachers and could call on the services of the Board's lecturers and a large number of volunteers. During that winter there were 3119 single talks or lectures, 102 short courses and 32 classes –

|  | Talks | Courses | Classes |
|---|---|---|---|
| Eastern Command | 1128 | 41 | 0 |
| A.A. Command | 1571 | 45 | 27 |
| R.A.F. | 186 | 4 | 0 |
| R.N. | 25 | 0 | 0 |
| A.T.S. | 188 | 12 | 5 |
| W.A.A.F. | 8 | 0 | 0 |
| U.S. Forces | 13 | 0 | 0[17] |

At the end of the War the Board accepted further responsibilities in this field. From August 1946 it organised the teaching of English to those members of the Free Polish Forces who had decided to settle in Great Britain.* This continued until there were no longer any students in East Anglia. In the following year the assistant secretary, Frank Bell, was responsible for an interesting experiment. There were eight camps for German and Austrian prisoners of war in the Cambridge area.† He first attempted to organise courses in the camps without success, but five of the camp commandants agreed to send students to a summer school in Cambridge which proved very successful.[18] It was repeated the following year when the prisoners sent a letter of thanks to the Board.[19]

Towards the end of the War the Board considered its future position in supplying adult education for the Forces, and in June 1944 it took the initiative in establishing the Regional Consultative Council for Adult Education in the Eastern Counties, which included representatives of local education authorities as well as the voluntary bodies. Because part of the Forces work was not of university standard the Board felt that it should be administered by the Regional Council.[20] However when the Eastern Counties Committee ceased its operations in July 1946, the Council was unable to take over, while the continuation of conscription maintained the demand. Therefore the Board, like departments of adult

* BEMS 48/19. Liverpool and Edinburgh ran similar courses (Kelly, *Adult Education*, p. 372).

† At the end of the War prisoners were retained in England to work on farms and because it was impossible to repatriate them immediately to a devastated and divided country.

education in other universities, found it necessary to continue the work itself.[21] In 1948 the Central Advisory Council was replaced by a Central Committee for Adult Education in H.M. Forces. The abolition of conscription in 1957 reduced the demand and the Central Committee was succeeded by a smaller Committee for University Assistance to Adult Education. It was then agreed to limit activities to work of a university standard.[22] Cambridge still has considerable numbers of troops within its territory and provides residential and lecture courses for them – for which it receives a grant from the Ministry of Defence.

The other innovation of the War years was the residential course, but this was not to develop fully for another ten years. In September 1942 the Board organised a residential course on X-ray Crystallography in collaboration with the Cavendish Laboratory and the Department of Mineralogy.[23] Attendance was restricted to government and industrial scientists, but it was very successful. Because of the large number of unsuccessful applicants it was repeated in subsequent years.[24] In September 1947 a similar course on metal physics was begun, and in the following year four residential courses were held on different subjects.[25] For these courses the Board provided all the administration and the University Department concerned provided the lecturers and the equipment. This was obviously a good working arrangement, but it had two serious disadvantages for the Board. All the courses organised – and summer schools as well – had to be held during the Long Vacation which imposed a strain on its office staff.* In addition the accommodation available in the colleges was not always suitable for summer courses – particularly in the older colleges. Cheshunt College, built just before World War 1, was one of the Board's best centres, but it could not accommodate large numbers.

Before the War there had been a few adult education colleges – Ruskin College at Oxford for working men, Woodbrooke and Fircroft at Birmingham with a strong religious element, and Coleg Harlech in North Wales. The last, opened in 1927 by the efforts of Tom Jones,† was the only true extra-mural college. In 1947 the Ministry of Education recommended all adult education organisa-

---

* It should be remembered that Long Vacation at Cambridge includes an extra term for which a number of undergraduates go into residence.

† Tom Jones was assistant secretary at the Cabinet Office until 1930, when he became secretary of the Pilgrim Trust.

tions to consider the advantage of centres, either residential or non-residential.[26] This encouraged both local education authorities and extra-mural departments to acquire country houses, disused chapels and other redundant buildings for conversion.

The Board had anticipated the Ministry and as early as 1946 had arranged informal inspections of four mansions in East Anglia. It included the proposal in the first post-War plans for the University Grants Committee.[27] In 1948 the University promised Madingley Hall to be used by the Board out of term.[28] The village of Madingley, four miles north west of Cambridge, had been bought by the University, and the Hall, a Tudor mansion with Victorian additions, had been included in the sale. During term it was to be a hall of residence for post-graduate students and out of term it was to be an extra-mural centre. The Board had proposed five types of short course which could be held there:

(a) Courses ranging from a weekend to two or three weeks arising out of normal extra-mural classes . . .

(b) Refresher courses for tutors in adult education and for secondary school teachers in special subjects (especially for those whose training has been confined to Emergency Colleges), and courses for training potential tutors in adult education.

(c) Courses arranged in co-operation with other University Departments for persons engaged in industry, in government departments, etc. . . .

(d) Courses for foreign students . . . for prolonged specialist courses in such subjects as English law, English medical training and organisation, English government and administration, the English educational system, and English language and literature, etc.

(e) Short courses arranged with or for other organisations hitherto not so intimately connected with the Board's normal work. (There have recently been requests for residential work from, for example, members of Trade Unions, Local Government Officers, groups of Rotary Clubs, etc.)[29]

The Board also envisaged longer residential courses for small numbers of students. Since 1946 it had co-operated with the Y.M.C.A. in organising such courses for 'men from the managerial and technical side of industry'.* Between six and twelve men nominated and financed by their employers spent a term at Cambridge attending some ordinary university lectures, some special lectures and discussion groups.[30]

---

* Until 1957 these courses were at Cheshunt College. When that college left Bateman Street, the courses were transferred to University College.

Although Madingley Hall was a considerable distance from Cambridge and money was needed to rebuild the kitchens and to adapt the courtyard to provide additional rooms, nevertheless, it was an important addition to the Board's work. Unfortunately work on the Hall was not completed until Michaelmas 1951 and to the Board's great regret it was not available during the Long Vacation of that year.[31] Frank Bell was resident M.A. during Canon Raven's wardenship of Madingley Hall and from 1954 to 1961 the chairman of the Board, F. R. Salter, was Warden.* In 1952 the Board used the Hall to the limits of its capacity. It was usually possible to organise just over twenty courses each year with a total average attendance of 500 or 600.† The courses included industrial and economic history for students drawn from industry or trade unions, current affairs for members of the British and American Forces, conferences for full-time and part-time staff of the Board and other adult education organisations, and meetings of the Local Centres Union. The enthusiasm which all these activities inspired – particularly from local extension societies[32] – increased the pressure on accommodation at Madingley. In 1952 the Board began to organise residential courses on behalf of the British Council and the Faculty of Law; in 1953 courses for apprentices and junior managers; in 1957 a colloquium for mathematics teachers and in 1963 a Russian language course.[33] The increase in the number of courses made it necessary to hold some elsewhere and it was fortunate that the building of two new colleges – Fitzwilliam and Churchill – enabled the Board to do so easily. From 1965 it was possible to hold courses for magistrates and from 1967 long courses in general studies for employees of I.B.M.[34] In addition residential courses are now held in other parts of East Anglia (Flatford Mill, Ipswich and Keswick Hall, Norwich), in other parts of England (Brathay Field Study Centre in the Lake District and Dartington Hall in Devon), and in other countries (Malta and Cyprus).‡ Even now it is necessary to restrict the number of residential courses to prevent the Board's resources being over-extended during Long Vacation.

* F. R. Salter was tutor of the first Norwich class in 1912. He was a member of the Syndicate from 1914 to 1924 and of the Board from 1924 to 1957. He died in 1968 (*44th Annual Report*, 1968).　　　　† See Table 3.
‡ The Syndicate had accepted a request for a course in Valparaiso in 1884 (Local Exam. Syndicate minute book 2, 9 May and 11 June 1884), and organised courses on the Riviera in 1897 (BEMS 55/27).

Although World War 2 brought the Summer Meetings to an end, one aspect of the work still continues in Cambridge as the Vacation Course for Foreign Students. As early as 1896 considerable numbers of foreign students were attending summer meetings to improve their English as well as their general knowledge. One student from Bavaria in that year proposed more advertising in Europe, and 'lectures on the English language, especially on phonetics and composition'.* By 1904 the Summer Meeting included lectures in French and courses on 'The Use of the Phonograph in the Teaching of Modern Languages' and 'The Teaching of English in German Secondary Schools'.[35] By 1910 nearly a third of the students attending came from outside the British Isles.†

Gradually more and more courses were provided for foreign students until it became possible to organise separate meetings. The first of these was in the Long Vacation of 1925 (when the Summer Meeting was at Oxford) on 'Contemporary England: its Language, Literature, Institutions and Music'.[36] Special classes in conversational English were included and certificates of competence awarded. Under the Board they were held in alternate years – when the Summer Meeting was at Oxford – and numbers rose to the maximum of 300. They were very successful: 'It was the first visit to England for most of these students. To begin with the slender bond binding them together was a common desire to be able to speak and understand the English tongue. But the number was large enough for each individual to discover in the crowd a kindred spirit or spirits . . .'[37]

From time to time there had been proposals that the Board might also run courses for American students and when Dr Cranage spent a term in the United States he had discussions about this at the University of Pittsburgh.‡ A report he presented to the Board in 1928 showed that the Foreign Students Course was not only successful but also profitable, and suggested an American Students' Course of three or four weeks on similar lines for graduates or teachers.§ This was first held in the summer of 1929

* *University Extension Journal*, vol. 2, p. 23. See also p. 37. In 1896 69 foreign students were present (*23rd Annual Report of Syndicate*, p. 4).
† See the tables for 1904–10 in *37th Annual Report of Syndicate*, p. 6.
‡ These are not mentioned in his description of the tour (*Not Only a Dean*, p. 145). See BEMS 4/2, 122.
§ BEMS 4/2, 122. A fee of £4 was to be charged for the course.

and other English-speaking students were allowed to attend.[38] It was decided to continue despite the international financial crisis and the Secretary was sent on a tour of the United States and Canada to explore the possibilities of attracting more students. In 1933 he made a second trip across the Atlantic and in 1938 a tour of the eastern United States.[39] On this occasion he discussed a longer course of six weeks for which the students could claim 'credits' towards a degree. Although one such course was held in the summer of 1939, most Americans and Canadians attending Cambridge courses in the next seven years were members of the Armed Forces.

Courses for foreign students did not cease during the War, but they were intended for troops or refugees. By 1949 when the Board was ready to return to normal courses, the University was considering its international activities.* Although the U.S. Educational Commission asked the Board to provide a summer school in 1952, most American universities were' providing their own teaching facilities in Great Britain where they thought them necessary.† But there was clearly a need for more assistance to the students of other countries. Frank Bell submitted a memorandum to the Board in 1949 suggesting that the University should make the Board its agent for this work.‡ Unfortunately the University was unable to develop its international activities to the extent originally envisaged and there has been no central co-ordination of the work in the way he proposed. The Board's Vacation Course is now held every year for a month. Lectures and seminars are given on British institutions and literature, classes are held in conversational English, and various excursions, visits and social activities are included. In 1964 numbers had to be restricted because the Examination Halls and Arts School ceased to be available,[40] but this problem was solved by a move to the Sidgwick Avenue site. Vacation Courses are widely advertised in Europe and very popular.

In 1946 the scheme for bursaries for adult students was revived and the first students to arrive in Cambridge were those who had been accepted in 1939, but who could not take up their bursaries because of the War. There were now many other sources of

* BEMS 48/22.
† BEMS 48/24. In 1948 it had been decided not to hold American Summer Schools (*24th Annual Report*).
‡ BEMS 48/22. In 1954 he resigned to found the Bell School of Languages at Cambridge.

scholarships for mature students – Senior State Scholarships and Ex-Servicemen's Scholarships in particular – and the Board through its Bursaries Committee accepted a supervisory responsibility for all these students without accepting the financial responsibility.[41] Despite the number of alternatives the number of applicants for the Board's bursaries was not reduced – 178 applied in 1947. The Board also now had money available – maintenance could now be paid for wives, children and other dependants. Special grants were made to a student whose Ministry scholarship had been reduced and to another who had been elected President of the Cambridge Union.[42] At the peak of operation in 1951 the Board was responsible for eighteen students – six with bursaries and twelve with other grants.[43]

In 1950 the Oxford Delegacy suggested a conference with Cambridge about the award of bursaries. The Delegacy was dissatisfied with the selection procedure and the examination results obtained, and proposed a special entrance examination which the Board thought unnecessary.* However, soon afterwards the number of Cambridge applicants began to fall and in 1952 the Bursaries Committee expressed doubts about the future of the scheme. They suggested a return to the original intention of a one year residence not leading to a degree and thought it could be based on Madingley.[44] This did not prove very popular or successful and was soon abandoned. From 1964 the number of applicants whom the Board thought suitable fell away rapidly. In 1965 and subsequent years only one bursary could be awarded and in 1969 and 1970 none. A survey showed that there were increased opportunities to obtain central or local government grants so that 'there are now very few mature students of degree calibre who cannot obtain a university place combined with a substantial grant'.† The Board therefore amended the scheme to make grants of £100 to assist mature students who already had a scholarship.

Since 1945 the work of the Board has been increasingly dependent on grants from the central government.‡ The Education Act of 1944 made fresh regulations for adult education necessary.§

* BEMS 8/2, 38, 42, 46. BEMS 48/23. The Delegacy was also concerned about the large number of students who chose Oxbridge.

† Minutes of the Bursaries Committee, 7 January 1971, and appendices.

‡ While some is received direct from the Treasury, other grants come through the University or through local education authorities.

§ Act 7 and 8 Geo. VI, c. 31. Grant Regulations no. 6 of 1946.

When issued in 1946 these swept away the old distinctions and defined adult education as:

(a) any establishment for affording to persons over compulsory school age full-time or part-time education or leisure-time occupation in organised cultural training and recreation activities; and

(b) any course of instruction forming part only of the work of an establishment as aforesaid or conducted separately.

It was to be provided by a university or university college, an approved association or a joint committee – all of which were now to be known as 'responsible bodies'. The Minister of Education was given a wide discretion in approving adult education courses and was to give a grant which 'shall not be less than three-quarters of the approved cost'.*

The new regulations and financial difficulties caused by the War led an increasing number of extension centres to accept government aid.† From 1940 onwards the Board was willing to take the Board of Education grant for the course and pay the lecturer's fee. This freed the local centre from a large financial burden and enabled them to be more adventurous in their choice of subjects. At Derby this reduced the Extension Society's payments from £105 3s 6d a session in 1925–6 to £36 19s 9d in 1945–6.‡ The Colchester Extension Society found its annual loss of £10 (met by a grant from the Education Committee) converted into a small profit by the same means.§ The local centre which adopted this method now had only to meet the cost of hiring a hall and advertising the course. Unfortunately the disappearance of financial problems occasionally led to the disappearance of a sense of financial responsibility. In 1952, 1956 and 1971 the Board thought it advisable to ask for a small payment for each course.‖

Although the increased grants have helped the Board to provide more courses where they were most needed and has enabled extension societies and W.E.A. branches to remain solvent in a time of

* The regulations of 1946 have been frequently amended, most recently in 1969 (Statutory Instrument no. 403), but still retain the same flexibility.

† Tutorial and preparatory classes were automatically supported by grants.

‡ Derby Univ. Ext. Soc., *Annual Reports*, 1925–6 and 1945–6. In the former year it paid £69 10s 6d for the lecturer, £21 10s for the hall and £14 3s for advertising.

§ Colchester Univ. Ext. Soc., account book, 1919–55.

‖ BEMS 4/3, 60. Lowestoft Literary and Scientific Assoc., minute book 2. BEMS – Ely Univ. Ext. Committee minute book, 1938–58, f. 56.

inflation, they have also had serious disadvantages. The amount of money available and the fees charged to students have become a matter of political and economic interest. In 1953, for example, the Conservative Government proposed a cut of 10 per cent following a standstill on grants in the previous year.* Despite Winston Churchill's lack of interest in educational matters, the storm of protest persuaded him to intervene and shortly afterwards the Ashby Committee was appointed to investigate the organisation and finance of adult education.† Although the Committee found a situation of considerable complexity, it felt that it worked well on the whole. It recommended an increase in finance with an easing of the financial regulations, but gave the Minister greater control over policy.[45]

Subsequent attacks on adult education have been limited to financial restrictions and are possibly less politically motivated. In 1963 the Robbins Committee on Higher Education recommended increased grants for adult education, but in the following year the newly-elected Labour Government proposed to hold grants at the existing level despite inflation. The Universities Council for Adult Education published its protest in May 1964, showing how little adult education cost in comparison with other government spending.[46] A small increase in the grant was permitted, but subse-quently there were further restrictions on grants and an increase in the fees paid by students. The Cambridge Board is less dependent on these grants than similar departments in other universities. Some of its courses are self-financing and may even make a contribution to the Board's funds. Nevertheless the uncertainties of the grants make it difficult to plan ahead. In 1971–2, for example, financial restrictions made it necessary to reduce the meetings in sessional classes from 24 to 21.[47]

On the other hand government grants have enabled the Board to treat its full-time staff more fairly. Although the Board (and before 1924 the Syndicate) had always been concerned to give some security to lecturers and tutors who depended on them for a living, it was impossible to give them security of tenure when

* At this time the number of Cambridge courses and classes was reduced by about one third. See *Report of the Universities Council for Adult Education, 1952–3*, pp. 5–8, 32–5.

† BEMS 4/3, 58. Kelly, *Adult Education*, p. 340. I am also indebted to Mr Hardman, then an Opposition M.P., for his reminiscences of this incident.

finance was uncertain.* Several schemes to provide some security and a pension failed for this reason. In 1939 the Board employed two assistant secretaries, two James Stuart lecturers with a guaranteed minimum stipend, and four resident tutors. Although each of these posts was more or less permanent, the holders were only appointed for a few years at a time. The rest of the full-time lecturers, about fifteen in all, had no certainty of employment from year to year.

The War and increased grants changed the staff position completely. Extra staff had to be appointed at Stuart House and in 1941 the first women clerks appeared.† Extra lecturers had to be found both to replace those on War service and to cope with the increased demand for courses. After the War the place of the full-time lecturer dependent on fees alone was taken by a greatly increased number of salaried lecturers and an increased number of part-time lecturers. There was no longer the need for the old-style extension lecturer. While Cambridge centres had been scattered across England, the part-time lecturer was unsuitable, even if such people had been available. After the War the Board had a compact area in East Anglia and the East Midlands, while there were far more people available with suitable qualifications and time to spare for some lecturing. The increase in the number of museum curators and archivists alone provided many more part-time lecturers in subjects which were increasingly popular.‡ It also became easier to employ intra-mural lecturers, although few were willing to help outside residential courses. As a result of these changes the Board abandoned its former requirement of trial lectures and the restriction on the number of lecturers in each subject because they were no longer relevant.

By 1947 the Board had sixteen permanent teaching staff. Resident tutors were appointed for all counties and there were staff tutors for a wide variety of subjects. The number of staff at Stuart House also increased, and most of the assistant secretaries

* Since many university and college appointments at Cambridge are for limited periods, it was also difficult to convince the University that extra-mural lecturers should be treated differently.

† It was not until 1969 that the first woman assistant secretary was appointed.

‡ In the Board's area record offices were established in Bedfordshire (1913) and Essex (1938) before the War, in East and West Suffolk (1950 and 1955) and Northamptonshire (1952) soon after the War, and in Cambridgeshire, Huntingdonshire and Norfolk more recently. All have co-operated in organising courses in local history.

appointed were able to do some teaching. In 1945 David Hardman resigned his post of assistant secretary on his election to Parliament.* G. H. Pateman was joined by Frank Bell. In 1950 a third assistant secretary, John Andrew, was appointed.† In 1955 Mr Pateman, who had been an assistant secretary since 1935 and secretary of the W.E.A. district before that, retired, and Mr Bell resigned.‡ They were replaced by Vivian Ramsbottom, who had previously been a resident tutor, and Dennis Raymond who resigned three years later. He was succeeded by Alan Stripp. In 1969 Miss Elizabeth Harrison was appointed to the new post of fourth assistant secretary.

The post-War years also saw the disappearance of many former restrictions and the problems associated with them. No longer was there a division between courses provided by the Local Centres Committee and classes provided by the Joint Committee with the W.E.A.§ Tutorial classes could now be urged on extension societies and W.E.A. branches could take extension courses without problems arising.‖ Often the lecturer and the students may not have been aware which committee had sponsored the course. In 1964 the Local Centres Committee disappeared after 91 years existence in different forms,¶ and the Joint Committee was only retained because it provided a means of closer co-operation with the W.E.A. This was a time when the University was attempting to streamline its administration, and Geoffrey Hickson suggested that the Board should obtain a new constitution after 40 years.** Membership of the Board was reduced from 30 to 18, and its standing committees were reduced to four – Finance and General Purposes, Bursaries, Appointments and the Joint Committee.†† There was no need to change the Board's powers which had proved themselves sufficiently wise and flexible since 1924.

* Mr Hardman still lectures for the Board 48 years after his first course – given at Wellingborough.

† Mr Andrew had been a mature student under the Board's supervision at St Catherine's. He is now Secretary of the Board.

‡ Mr Pateman died in 1968 (*44th Annual Report*). Mr Bell founded the Bell School of Languages and continued some of the work he had done for the Board.

§ The Local Lectures Committee and the Tutorial Classes Committee which had been renamed in 1946.

‖ *32nd Annual Report*, 1956. Colchester Univ. Ext. Soc., minute book 1941–55, ff. 50 and 65.

¶ It was the descendant of James Stuart's original Syndicate, which became the Local Lectures Committee in 1878.

** Interview with Mr Hickson in 1971. Graces 1 and 2 of 26 June 1964.

†† The staff meeting, for which minutes are also kept, is not a committee of the Board.

The disappearance of distinctions in the organisation of courses and classes produced some changes in the subjects studied. History, and particularly local history, became more popular with tutorial classes because a three year period of study is ideal for the research involved.* Economics and political subjects became less popular with tutorial classes, but more popular with extension courses. These subjects and psychology (which followed the same shift) can often be better studied in terminal or sessional courses. The only other change in the extension field was a considerable decline in English Literature which had been very popular from 1873 onwards.† The disappearance of the separate class and the examinations also allowed extension courses to be more flexible. Music courses could include a visit from the Phoenix String Orchestra conducted by an assistant secretary.‡ Local history courses could be held in record offices so that members could pursue their own research. Some of the results of this co-operative research have been published locally.§ The latest venture of the Board in this field is a certificate in local studies to be awarded at the end of a four year course.‖ In a changing world only the students seem to remain the same. The register of a Bedford sessional class in Comparative Religion in 1966 is little different from the register of a tutorial class in Social and Industrial History in 1920. The same kind of people‡came from the same areas of the town (after allowance is made for its growth).¶ The only noticeable difference is the increased number of husbands and wives who attend.

As the Board continued to surrender its centres outside the Eastern Counties and to transfer the outlying parts of its territory to other universities, so the possibilities of co-operation with other departments of adult education increased.** Glasgow in 1959,

* An increased interest in local history, for which the sources were now more readily available, was also the national trend (*Report of Universities Council on Adult Education*, 1951–2, p. 14).
† See Table 6.
‡ E.g. Wellingborough University Extension Society, annual report, 1961. The assistant secretary was Alan Stripp.
§ E.g. a tutorial class at Cottenham Village College published *Charity School to Village College* in 1968.
‖ Board circular of 29 May 1968.
¶ See Appendix; information from BEMS 24/9 and 12.
** Derby will be surrendered to Nottingham in 1973. Northamptonshire was transferred to Leicester in 1971. Cambridge will have no centres outside East Anglia after 1973.

Nottingham and Exeter in 1970 held joint residential courses with the Board.[48] The possibilities of friction have disappeared. In 1962 the Board reaffirmed the position which the Syndicate had taken in 1877 when it withdrew from Leeds in favour of the Yorkshire College.[49] In 1965 the Board recognised the interest of the Universities of Essex and East Anglia in their field by co-opting representatives of each to the Board.[50]

In September 1967 Geoffrey Fletcher Hickson retired from the post of Secretary to the Board which he had held since 1928, and was succeeded by his deputy, John Andrew. In a traditional gesture of goodwill Mr Hickson's portrait was painted and hung in the Boardroom with those of his predecessors.* Less formally, and particularly appropriately, some of his colleagues published a *festschrift* in his honour.[51] It was particularly appropriate because he was the first retiring Secretary with enough professional colleagues in Cambridge to make it possible, because the essays were mainly historical in tribute to an historian who became an administrator, and because it was devoted to East Anglia where the Board's work now lay. During his career the retiring Secretary had seen more changes in adult education than his predecessors. Under his guidance the Board had adapted its policies to suit the changing circumstances. He saw clearly what needed to be done and even though the Board was not always so far-sighted they usually followed his advice in the end. It could be said of all the secretaries, as it was said of Dr Cranage, that they were able to convince others of the correctness of their views. It could also be said of most of the secretaries that they saw what was needed to further the work of University Extension and they planned accordingly. Geoffrey Hickson will not be the last Secretary of the Board of Extra-Mural Studies with these valuable attributes.

* Part of the cost was met by donations from local centres (Wellingborough Univ. Ext. Soc., annual report, 1967).

TABLE 1. *Extension courses*

Where possible these figures are taken from the *Annual Reports* of the Syndicate and the Board. In the early years few statistics were kept and those available may not be completely accurate. The first set of figures for each year are for the Lent term and the second for the Michaelmas term.

| | | | Number of students attending | | |
|---|---|---|---|---|---|
| | Centres | Courses | Lecture | Class | Examina-tion |
| 1873 | 0 | 0 | 0 | 0 | 0 |
| | 3 | 8 | n.a. | n.a. | n.a. |
| 1874 | 7 | 19 | n.a. | n.a. | n.a. |
| | 11 | 24 | n.a. | n.a. | n.a. |
| 1875 | 16 | 30 | n.a. | n.a. | n.a. |
| | 33 | 78 | n.a. | n.a. | 744 |
| 1876 | 26 | 50 | n.a. | n.a. | 980 |
| | 31 | 50 | 5002 | 3170 | 634 |
| 1877 | 22 | 38 | 2509 | 1663 | 441 |
| | 20 | 36 | 3545 | 2520 | 569 |
| 1878 | 16 | 31 | 2395 | 1618 | 519 |
| | 17 | 32 | 3043 | 2029 ⎫ | 556 |
| 1879 | 15 | 24 | 1877 | 1172 ⎭ | |
| | 19 | 29 | 3570 | 2704 | 411 |
| 1880 | 13 | 22 | 1439 | 1044 | 283 |
| | 13 | 18 | 2510 | 1465 | 311 |
| 1881 | 13 | 19 | 1859 | 1159 | 262 |
| | 17 | 23 | 2006 | 1226 | 303 |
| 1882 | 12 | 20 | 1400 | 874 | 199 |
| | 30 | 37 | 3800 | 2027 | 519 |
| 1883 | 19 | 26 | 2474 | 1411 | 387 |
| | 30 | 39 | 4522 | 2765 | 660 |
| 1884 | 21 | 27 | 3356 | 2018 | 515 |
| | 33 | 41 | 4597 | 2619 | 709 |
| 1885 | 24 | 32 | 2662 | 1533 | 384 |
| | 36 | 43 | 4710 | 2429 | 756 |
| 1886 | 25 | 37 | 3847 | 1951 | 622 |
| | 45 | 57 | 6430 | 2976 | 835 |
| 1887 | 29 | 43 | 4064 | 1844 | 506 |
| | 45 | 60 | 5980 | 2948 | 962 |
| 1888 | 29 | 38 | 3529 | 1705 | 569 |
| | 37 | 52 | 5677 | 2515 | 793 |
| 1889 | 25 | 37 | 3618 | 1961 | 627 |
| | 47 | 69 | 6247 | 3031 | 944 |
| 1890 | 38 | 56 | 5054 | 2313 | 790 |
| | 50 | 71 | 6398 | 2850 | 876 |
| 1891 | 44 | 64 | 4549 | 2066 | 671 |
| | 140 | 161 | 9701 | 4347 | 1029 |
| 1892 | 143 | 155 | 8584 | 4032 | 1203 |
| | 102 | 113 | 7736 | 3756 | 924 |
| 1893 | 91 | 103 | 7882 | 3186 | 834 |
| | 66 | 78 | 5638 | 2125 | 676 |

TABLE 1 (*cont.*)

| | | | Number of students attending | | |
|---|---|---|---|---|---|
| | Centres | Courses | Lecture | Class | Examina-tion |
| 1894 | 46 | 55 | 4814 | 2092 | 739 |
| | 44 | 54 | 6552 | 2264 | 633 |
| 1895 | 31 | 42 | 3786 | 1690 | 659 |
| | 37 | 48 | 4945 | 1955 | 535 |
| 1896 | 39 | 49 | 4236 | 1741 | 554 |
| | 32 | 40 | 3496 | 1305 | 380 |
| 1897 | 41 | 46 | 5000 | 1287 | 363 |
| | 38 | 45 | 6153 | 1265 | 252 |
| 1898 | 50 | 58 | 8277 | 1390 | 344 |
| | 54 | 57 | 5569 | 1656 | 310 |
| 1899 | 51 | 62 | 5426 | 1451 | 438 |
| | 49 | 58 | 5581 | 1414 | 346 |
| 1900 | 44 | 54 | 3872 | 1019 | 332 |
| | 42 | 48 | 5250 | 1502 | 174 |
| 1901 | 45 | 52 | 4313 | 1306 | 487 |
| | 48 | 55 | 4721 | 1633 | 185 |
| 1902 | 40 | 48 | 4369 | 1467 | 524 |
| | 45 | 51 | 6088 | 1979 | 240 |
| 1903 | 51 | 58 | 6538 | 2164 | 573 |
| | 45 | 53 | 6080 | 1879 | 243 |
| 1904 | 56 | 65 | 6821 | 2286 | 779 |
| | 48 | 55 | 7040 | 1793 | 284 |
| 1905 | 51 | 62 | 5941 | 1622 | 674 |
| | 52 | 60 | 8036 | 1905 | 320 |
| 1906 | 35 | 44 | 5831 | 1087 | 245 |
| | 48 | 56 | 6828 | 1401 | 256 |
| 1907 | 46 | 55 | 6699 | 1735 | 342 |
| | 43 | 52 | 6461 | 1749 | 300 |
| 1908 | 42 | 50 | 5136 | 1328 | 183 |
| | 38 | 47 | 6421 | 1299 | 237 |
| 1909 | 41 | 49 | 5423 | 1194 | 207 |
| | 40 | 47 | 5038 | 1234 | 220 |
| 1910 | 43 | 50 | 5141 | 1498 | 158 |
| | 45 | 52 | 6499 | 1345 | 232 |
| 1911 | 34 | 39 | 4404 | 1350 | 115 |
| | 41 | 46 | 6232 | 1393 | 187 |
| 1912 | 33 | 35 | 3984 | 860 | 113 |
| | 43 | 49 | 6436 | 1416 | 181 |
| 1913 | 28 | 31 | 3052 | 844 | 87 |
| | 40 | 46 | 5015 | 959 | 155 |
| 1914 | 32 | 36 | 4418 | 1102 | 80 |
| | 35 | 40 | 4894 | 943 | 106 |
| 1915 | 20 | 21 | 2854 | 676 | 56 |
| | 26 | 29 | 4389 | 728 | 73 |
| 1916 | 13 | 15 | 1735 | 335 | 38 |
| | 23 | 26 | 3330 | 587 | 58 |

TABLE 1 (*cont.*)

| | | | Number of students attending | | |
|---|---|---|---|---|---|
| | Centres | Courses | Lecture | Class | Examina-tion |
| 1917 | 13 | 14 | 2020 | 571 | 36 |
| | 20 | 23 | 3522 | 793 | 61 |
| 1918 | 15 | 17 | 2754 | 544 | 15 |
| | 22 | 25 | 3526 | 608 | 51 |
| 1919 | 14 | 16 | 2515 | 765 | 23 |
| | 39 | 48 | 6163 | 1332 | 100 |
| 1920 | 26 | 29 | 4213 | 865 | 56 |
| | 45 | 54 | 7518 | 1895 | 204 |
| 1921 | 33 | 37 | 5382 | 1430 | 56 |
| | 47 | 56 | 7436 | 1922 | 128 |
| 1922 | 32 | 36 | 4285 | 1105 | 48 |
| | 42 | 47 | 6209 | 1458 | 158 |
| 1923 | 19 | 23 | 3379 | 692 | 34 |
| | 37 | 41 | 5093 | 1302 | 143 |
| 1924 | 24 | 26 | 3401 | 945 | 65 |
| | 41 | 46 | 4527 | 1199 | 117 |
| 1925 | 27 | 29 | 3160 | 1188 | 55 |
| | 34 | 40 | 4242 | 1408 | 169 |
| 1926 | 22 | 23 | 2426 | 556 | 92 |
| | 32 | 35 | 4241 | 1304 | 105 |
| 1927 | 24 | 26 | 2279 | 670 | 77 |
| | 33 | 37 | 4362 | 1285 | 141 |
| 1928 | 20 | 21 | 2485 | 723 | 41 |
| | 34 | 38 | 4076 | 1329 | 157 |
| 1929 | 21 | 23 | 2430 | 768 | 38 |
| | 33 | 37 | 3891 | 1008 | 104 |
| 1930 | 16 | 17 | 2242 | 755 | 23 |
| | 30 | 35 | 3709 | 1171 | 93 |
| 1931 | 16 | 17 | 1610 | 597 | 35 |
| | 28 | 33 | 2988 | 893 | 64 |
| 1932 | 19 | 20 | 1632 | 587 | 37 |
| | 27 | 29 | 2811 | 945 | 69 |
| 1933 | 16 | 16 | 1244 | 437 | 31 |
| | 27 | 30 | 2697 | 887 | 71 |
| 1934 | 18 | 18 | 1525 | 509 | 14 |
| | 27 | 31 | 3054 | 987 | 52 |
| 1935 | 16 | 16 | 1366 | 407 | 19 |
| | 37 | 40 | 3045 | 1396 | 35 |
| 1936 | 10 | 10 | 822 | 301 | 10 |
| | 36 | 38 | 2815 | 1254 | 48 |
| 1937 | 14 | 15 | 953 | 325 | 4 |
| | 32 | 35 | 2730 | 1308 | 45 |
| 1938 | 15 | 15 | 855 | 533 | 13 |
| | 29 | 31 | 2209 | 1148 | 40 |
| 1939 | 15 | 16 | 1269 | 812 | 11 |
| | 9 | 9 | 1117 | 778 | 5 |

TABLE 1 (*cont.*)

| | Centres | Courses | Number of students attending | | |
|---|---|---|---|---|---|
| | | | Lecture | Class | Examina-tion |
| 1940 | 27 | 28 | 1741 | 1455 | 6 |
| | 25 | 25 | 1448 | 990 | 13 |
| 1941 | 21 | 22 | 1202 | 979 | 4 |
| | 38 | 38 | 2388 | 1964 | 11 |
| 1942 | 42 | 45 | 2158 | 1905 | 4 |
| | 36 | 38 | 1773 | 1614 | 9 |
| 1943 | 33 | 36 | 1629 | 1512 | 2 |
| | 40 | 42 | 1983 | 1607 | 2 |
| 1944 | 22 | 39 | 1762 | 1643 | 4 |
| | 39 | 43 | 2255 | 1972 | 6 |
| 1945 | 32 | 50 | 2392 | 2254 | 3 |
| | 32 | 35 | 1504 | 1307 | 0 |
| 1946 | 30 | 34 | 1306 | 1211 | |
| | 40 | 45 | 1787 | 1502 | |
| 1947 | 28 | 39 | 1256 | 977 | |
| | 40 | 44 | 1758 | 1673 | |
| 1948 | 26 | 31 | 1218 | 1133 | |
| | 35 | 38 | 1407 | 1249 | |
| 1949 | 26 | 36 | 1167 | 1116 | |
| | 36 | 40 | 1649 | 1408 | |
| 1950 | 28 | 35 | 1016 | 886 | |
| | 32 | 34 | 1190 | 1139 | |
| 1951 | 19 | 21 | 1058 | 571 | |
| | 28 | 31 | 1054 | 992 | |
| 1952 | 18 | 20 | 669 | 664 | |
| | 27 | 27 | 1135 | 990 | |
| 1953 | 19 | 19 | 610 | 584 | |
| | 22 | 22 | 896 | 790 | |
| 1954 | 19 | 21 | 725 | 670 | |
| | 24 | 25 | 1005 | 967 | |
| 1955 | 14 | 15 | 532 | 530 | |
| | 24 | 25 | 876 | 820 | |
| 1956 | 22 | 23 | 1131 | 859 | |
| | 29 | 34 | 1251 | 1221 | |
| 1957 | 18 | 19 | 1040 | 998 | |
| | 30 | 36 | 1309 | 1280 | |
| 1958 | 18 | 21 | 777 | 763 | |
| | 26 | 34 | 1150 | 1119 | |
| 1959 | 14 | 19 | 696 | 641 | |
| | 22 | 28 | 899 | 877 | |
| 1960 | 14 | 16 | 788 | 787 | |
| | 20 | 25 | 890 | 878 | |
| 1961 | 14 | 18 | 833 | 828 | |
| | 21 | 24 | 793 | | |
| 1962 | 14 | 15 | 962 | | |
| | 20 | 24 | 918 | | |

TABLE 1 (*cont.*)

| | Centres | Courses | Number of students attending | | |
| --- | --- | --- | --- | --- | --- |
| | | | Lecture | Class | Examina-tion |
| 1963 | 16 | 17 | 693 | | |
| | 17 | 19 | 789 | | |
| 1964 | 12 | 13 | 472 | | |
| | 15 | 17 | 659 | | |
| 1965 | 11 | 12 | 553 | | |
| | 18 | 22 | 862 | | |
| 1966 | 11 | 12 | 422 | | |
| | 18 | 21 | 733 | | |
| 1967 | 13 | 13 | 342 | | |
| | 22 | 27 | 984 | | |
| 1968 | 15 | 16 | 438 | | |
| | 19 | 21 | 635 | | |
| 1969 | 21 | 22 | 853 | | |
| | 21 | 25 | 747 | | |
| 1970 | 16 | 17 | 500 | | |
| | 21 | 23 | 681 | | |
| 1971 | 16 | 16 | 485 | | |
| | 16 | 18 | 594 | | |
| 1972 | 13 | 18 | 703 | | |

## TABLE 2. *Tutorial and preparatory classes*

These statistics are taken from the *Annual Reports* on Classes until 1924 and from the *Annual Reports* thereafter. The first figure for students is the original enrolment and the second for those doing effective work.

| | Centres | Classes | Enrolled students | Effective students |
|---|---|---|---|---|
| 1909–10 | 3 | 3 | 90 | 78 |
| 1910–11 | 3 | 3 | 88 | 80 |
| 1911–12 | 3 | 3 | n.a. | 63 |
| 1912–13 | 4 | 4 | 117 | 84 |
| 1913–14 | 5 | 5 | 97 | 71 |
| 1914–15 | 5 | 5 | 95 | 46 |
| 1915–16 | 2 | 2 | 37 | 23 |
| 1916–17 | 2 | 2 | 44 | 26 |
| 1917–18 | 2 | 2 | 49 | 29 |
| 1918–19 | 2 | 2 | 40 | 28 |
| 1919–20 | 4 | 6 | 167 | 66 |
| 1920–1 | 8 | 10 | 262 | 140 |
| 1921–2 | 11 | 13 | 305 | 156 |
| 1922–3 | 11 | 12 | 279 | 164 |
| 1923–4 | 10 | 13 | 321 | 209 |
| 1924–5 | 14 | 19 | 456 | 282 |
| 1925–6 | 16 | 22 | 458 | 300 |
| 1926–7 | 15 | 23 | 543 | 334 |
| 1927–8 | 17 | 24 | 483 | 384 |
| 1928–9 | 16 | 27 | 534 | 416 |
| 1929–30 | 14 | 24 | 490 | 398 |
| 1930–1 | 16 | 30 | 620 | 482 |
| 1931–2 | 17 | 33 | 623 | 471 |
| 1932–3 | 19 | 33 | 608 | 480 |
| 1933–4 | 19 | 33 | 681 | 522 |
| 1934–5 | 22 | 35 | 659 | 499 |
| 1935–6 | 21 | 35 | 659 | 482 |
| 1936–7 | 20 | 33 | 613 | 458 |
| 1937–8 | 18 | 33 | 651 | 439 |
| 1938–9 | 16 | 28 | 475 | 347 |
| 1939–40 | 16 | 28 | 468 | 305 |
| 1940–1 | 15 | 24 | 385 | 269 |
| 1941–2 | 20 | 30 | 557 | 359 |
| 1942–3 | 21 | 28 | 523 | 358 |
| 1943–4 | 22 | 30 | 648 | 359 |
| 1944–5 | 43 | 63 | 1409 | 741 |
| 1945–6 | 50 | 73 | 1487 | 812 |
| 1946–7 | 49 | 74 | 1564 | 752 |
| 1947–8 | 43 | 56 | 1104 | 676 |
| 1948–9 | 48 | 66 | 1312 | 733 |
| 1949–50 | 44 | 68 | 1258 | 777 |
| 1950–1 | 68 | 99 | 1310 | 658 |
| 1951–2 | 50 | 72 | 1321 | 714 |
| 1952–3 | 52 | 71 | 1342 | 730 |
| 1953–4 | 52 | 72 | 1374 | 732 |

TABLE 2 (*cont.*)

|  | Centres | Classes | Enrolled students | Effective students |
|---|---|---|---|---|
| 1954–5 | 45 | 67 | 1292 | 720 |
| 1955–6 | 47 | 70 | 1330 | 721 |
| 1956–7 | 52 | 73 | 1449 | 755 |
| 1957–8 | 47 | 74 | 1471 | 707 |
| 1958–9 | 49 | 68 | 1455 | 679 |
| 1959–60 | 46 | 68 | 1474 | 782 |
| 1960–1 | 51 | 75 | 1454 | 754 |
| 1961–2 | 56 | 74 | 1465 | 739 |
| 1962–3 | 46 | 72 | 1488 | 740 |
| 1963–4 | 50 | 70 | 1516 | 751 |
| 1964–5 | 45 | 64 | 1167 | 690 |
| 1965–6 | 38 | 57 | 1052 | 542 |
| 1966–7 | 48 | 75 | 1378 | 780 |
| 1967–8 | 55 | 97 | 1820 | 1112 |
| 1968–9 | 49 | 104 | 1897 | 1041 |
| 1969–70 | 53 | 103 | 1921 | 1126 |
| 1970–1 | 53 | 104 | 2064 | 1464 |
| 1971–2 | 58 | 119 | 2380 | 1722 |

TABLE 3. *Residential courses*

The first number in each column represents the number of courses and the second the number of students. The figures are taken from the Board's *Annual Reports* and are for the academic year beginning in October.

|  | Total | | Madingley Hall | | Cambridge Colleges | | Elsewhere | |
|---|---|---|---|---|---|---|---|---|
| 1951–2 | 13 | 128 | 13 | 128 | | | | |
| 1952–3 | 21 | 478 | 21 | 478 | | | | |
| 1953–4 | 22 | 494 | 22 | 494 | | | | |
| 1954–5 | 24 | 517 | 24 | 517 | | | | |
| 1955–6 | 22 | 564 | 22 | 564 | | | | |
| 1956–7 | 20 | 493 | 20 | 493 | | | | |
| 1957–8 | 28 | 778 | 20 | 491 | 8 | 287 | | |
| 1958–9 | 28 | 916 | 19 | 499 | 9 | 417 | | |
| 1959–60 | 31 | 909 | 22 | 584 | 9 | 325 | | |
| 1960–1 | 34 | 943 | 22 | 623 | 12 | 320 | | |
| 1961–2 | 39 | 1107 | 27 | 750 | 12 | 357 | | |
| 1962–3 | 39 | 1259 | 22 | 614 | 17 | 645 | | |
| 1963–4 | 37 | 1270 | 19 | 584 | 18 | 686 | | |
| 1964–5 | 37 | 973 | 22 | 563 | 15 | 410 | | |
| 1965–6 | 45 | 1089 | 23 | 448 | 22 | 601 | | |
| 1966–7 | 44 | 1380 | 23 | 587 | 20 | 769 | 1 | 24 |
| 1967–8 | 50 | 1581 | 21 | 597 | 25 | 898 | 4 | 86 |
| 1968–9 | 52 | 1492 | 22 | 581 | 28 | 870 | 2 | 41 |
| 1969–70 | 67 | 1856 | 24 | 559 | 35 | 1085 | 8 | 212 |
| 1970–1 | 60 | 1490 | 23 | 495 | 32 | 907 | 5 | 88 |
| 1971–2 | 77 | 2099 | 25 | 617 | 36 | 1225 | 16 | 257 |

## TABLE 4. *Rural area courses*

The statistics for rural areas work are not well set out in the *Annual Reports*. Tutorial classes and terminal (extension) courses are listed both under those headings and under rural areas where appropriate. These have been omitted in these tables, because they appear in earlier tables. The information given varies from year to year, so that these statistics are not as accurate as those in earlier tables.

In 1945 the Rural Areas Committee ended its work. Its place in the *Annual Reports* is taken by a new category – Terminal Courses – which is not exactly the same. See Table 5.

|  | Centres | Courses | Effective students |
|---|---|---|---|
| 1927–8 | 11 | 15 | 271 |
| 1928–9 | 19 | 23 | 440 |
| 1929–30 | 14 | 16 | 381 |
| 1930–1 | 26 | 27 | 545 |
| 1931–2 | 27 | 32 | 770 |
| 1932–3 | 30 | 30 | 637 |
| 1933–4 | 33 | 34 | 818 |
| 1934–5 | 37 | 38 | 885 |
| 1935–6 | 40 | 41 | 691 |
| 1936–7 | 44 | 46 | 1080 |
| 1937–8 | 47 | 53 | 1223 |
| 1938–9 | 59 | 69 | 1572 |
| 1939–40 | 49 | 57 | 1324 |
| 1940–1 | 81 | 98 | 1965 |
| 1941–2 | 89 | 108 | 2322 |
| 1942–3 | 97 | 132 | 2847 |
| 1943–4 | 123 | 140 | 2984 |
| 1944–5 | 123 | 162 | 3251 |

## TABLE 5. *Terminal courses*

From Michaelmas 1945 the Rural Areas Committee ceased to exist and separate statistics were no longer kept for the work in rural areas. However the new category, Terminal Courses, which appears in the *Annual Reports* is to some extent a continuation of the rural areas short courses. There is very little difference between these and the extension courses, which had themselves earlier been known as 'terminal' and 'short' courses.

From 1945 to 1947 and since 1967 separate statistics for Lent and Michaelmas terms have not appeared in the Board's *Annual Reports*.

|  | Centres | Courses | Enrolled students | Effective students |
|---|---|---|---|---|
| 1945–6 | 61 | 77 | 1526 | 1047 |
| 1946–7 | 81 | 95 | 1840 | 1165 |
| 1947 (Mich) | 48 | 50 | 975 | 686 |
| 1948 | 42 | 43 | 787 | 604 |
|  | 32 | 32 | 557 | 397 |
| 1949 | 29 | 29 | 579 | 429 |
|  | 32 | 32 | 563 | 403 |
| 1950 | 29 | 29 | 490 | 401 |
|  | 24 | 24 | 381 | 283 |
| 1951 | 21 | 22 | 358 | 250 |
|  | 25 | 25 | 369 | 268 |
| 1952 | 22 | 23 | 358 | 285 |
|  | 25 | 25 | 375 | 271 |
| 1953 | 25 | 25 | 242 | 203 |
|  | 23 | 23 | 364 | 268 |
| 1954 | 10 | 10 | 136 | 108 |
|  | 15 | 16 | 206 | 157 |
| 1955 | 13 | 13 | 187 | 143 |
|  | 13 | 13 | 197 | 134 |
| 1956 | 8 | 8 | 125 | 94 |
|  | 11 | 11 | 169 | 138 |
| 1957 | 8 | 9 | 127 | 108 |
|  | 9 | 9 | 112 | 78 |
| 1958 | 11 | 11 | 197 | 145 |
|  | 16 | 16 | 268 | 190 |
| 1959 | 10 | 10 | 141 | 122 |
|  | 19 | 19 | 255 | 184 |
| 1960 | 14 | 14 | 219 | 169 |
|  | 15 | 21 | 283 | 201 |
| 1961 | 7 | 7 | 101 | 75 |
|  | 15 | 16 | 257 | 167 |
| 1962 | 3 | 3 | 47 | 30 |
|  | 20 | 20 | 277 | 176 |
| 1963 | 10 | 10 | 168 | 127 |
|  | 14 | 14 | 183 | 109 |
| 1964 | 7 | 9 | 161 | 111 |
|  | 14 | 15 | 203 | 139 |
| 1965 | 5 | 5 | 86 | 62 |
|  | 13 | 13 | 158 | 112 |

TABLE 5 (*cont.*)

|  | Centres | Courses | Enrolled students | Effective students |
|---|---|---|---|---|
| 1966 | 13 | 13 | 286 | 218 |
|  | 15 | 16 | 230 | 174 |
| 1967 (Lent) | 6 | 7 | 176 | 117 |
| 1967–8 | 21 | 25 | 401 | 291 |
| 1968–9 | 16 | 20 | 250 | 199 |
| 1969–70 | 10 | 12 | 256 | 165 |
| 1970–1 | 14 | 17 | 279 | 208 |
| 1971–2 | 10 | 13 | 177 | 132 |

## TABLE 6. *Subjects of courses and classes*

Subjects have been arranged in the same groups in each section. The approximate percentage is given in brackets after each figure for ease of comparison. It proved impossible to use the same years in each case.

| | Total | A | B | C | D | E | F | G | H |
|---|---|---|---|---|---|---|---|---|---|
| | | | | Extension Courses | | | | | |
| 1882–3 | 38 | 17 (47) | 10 (24) | 9 (23) | 1 (3) | 1 (3) | — | — | — |
| 1892–3 | 99 | 50 (50) | 28 (28) | 12 (12) | — | 9 (10) | — | — | — |
| 1902–3 | 89 | 21 (24) | 31 (34) | 30 (34) | 1 (1) | — | 6 (7) | — | — |
| 1912–13 | 60 | 5 (8) | 19 (31) | 19 (31) | 7 (12) | — | 9 (16) | 1 (2) | — |
| 1922–3 | 55 | 12 (20) | 8 (15) | 19 (40) | 1 (5) | — | 8 (15) | 7 (15) | — |
| 1932–3 | 37 | 3 (8) | 6 (17) | 13 (33) | 4 (11) | — | 6 (17) | 5 (14) | — |
| 1942–3 | 61 | 1 (1) | 6 (9) | 11 (19) | 29 (48) | — | 11 (19) | 3 (4) | — |
| 1952–3 | 108 | 2 (2) | 40 (37) | 9 (8) | 39 (36) | — | 16 (15) | 2 (2) | — |
| 1962–3 | 108 | 6 (5) | 28 (29) | 10 (9) | 35 (33) | — | 15 (14) | 4 (3) | 10 (9) |
| | | | | Tutorial Classes | | | | | |
| 1909–10 | 3 | — | — | 1 (33) | 2 (67) | — | — | — | — |
| 1919–20 | 6 | — | 4 (66) | 1 (17) | 1 (17) | — | — | — | — |
| 1929–30 | 24 | 3 (13) | 3 (13) | 4 (18) | 7 (28) | 2 (9) | — | — | 5 (19) |
| 1939–40 | 28 | 2 (7) | 4 (14) | 5 (16) | 7 (23) | — | — | — | 10 (40) |
| 1949–50 | 68 | 4 (6) | 13 (19) | 8 (12) | 21 (30) | — | 6 (9) | 4 (6) | 12 (18) |
| 1959–60 | 68 | 4 (6) | 22 (39) | 9 (13) | 15 (22) | — | 6 (9) | 2 (3) | 10 (15) |
| 1969–70 | 103 | 11 (10) | 44 (43) | 8 (8) | 20 (19) | — | 8 (8) | 3 (3) | 9 (9) |
| | | | | Residential Courses | | | | | |
| | | | | | | Misc. | | | |
| 1951–2 | 13 | 1 (8) | 2 (15) | 3 (23) | 2 (15) | 1 (8) | — | — | 4 (31) |
| 1956–7 | 20 | — | 5 (25) | 3 (17) | 12 (58) | — | — | — | — |
| 1961–2 | 39 | 1 (2) | 6 (16) | 4 (10) | 19 (50) | 8 (20) | 1 (2) | — | — |
| 1966–7 | 44 | 1 (2) | 9 (21) | 4 (9) | 18 (43) | 10 (21) | 1 (2) | 1 (2) | — |

A – Science
B – History, including archives
C – Literature, mainly English
D – Economics, Social and Political Theory
E – Geography
F – Art and Architecture
G – Music
H – Psychology and Philosophy

TABLE 7. *Adult students*

These statistics are taken from the information assembled by Mr Pateman in 1938 and 1949 (BEMS 56/8). Because some of the replies to his questionnaires were incomplete, totals differ slightly. 44 students are listed between 1923 and 1937, and 19 between 1946 and 1949.

|  | 1923–37 | 1946–9 |
|---|---|---|
| *Previous occupations* | | |
| Skilled Manual | 8 | 2 |
| Unskilled Manual | 11 | 2 |
| Clerk | 8 | 3 |
| Teacher | 6 | 1 |
| Civil Servant | 4 | 5 |
| Shopworker | 2 | 1 |
| Journalist | 1 | 1 |
| Insurance Agent | 1 | 0 |
| Organist | 1 | 0 |
| Adult Education | 0 | 1 |
| Nurse | 0 | 1 |
| *New occupations* | | |
| Adult Education | 10 | 2 |
| Return to old occupation | 7 | 0 |
| Social Work | 6 | 1 |
| Civil Service | 4 | 1 |
| University | 2 | 0 |
| Died | 2 | 0 |
| C.W.S. | 1 | 0 |
| Teacher | 0 | 1 |
| *Form of adult education* | | |
| Tutorial Classes | 30 | 13 |
| Extension Courses | 3 | 1 |
| Both | 6 | 2 |
| *Degree obtained* | | |
| Doctorate | 2 | 0 |
| First Degree | 27 | 7 |
| No Degree | 9 | 0 |
| Not yet taken | 6 | 12 |
| *Age on entry* | | |
| 20 | 1 | 0 |
| 21–30 | 32 | 12 |
| 31–40 | 10 | 6 |
| 41–50 | 1 | 1 |

## TABLE 8. *A lecturer's courses*

In March 1903 A. Hamilton Thompson listed all the courses he had given for the Syndicate since 1897 (BEMS 55/27).

| | | | |
|---|---|---|---|
| 1897 | Mich. Term | Braintree<br>Harpenden<br>Hertford<br>St Ives (Hunts.)<br>Stevenage | } Victorian Poets and Novelists |
| 1898 | Lent Term | Grantham | Renaissance |
| | | Norwich<br>Hunstanton<br>Diss | } Architecture |
| 1898 | Mich. Term | Southport | Shakespeare |
| | | Huntingdon<br>Ipswich<br>Retford | } Architecture |
| 1899 | Lent Term | Saffron Walden<br>Braintree (class) | } Renaissance |
| | | Bury St Edmunds<br>Derby<br>Tewkesbury | Architecture<br>Victorian Poets<br>Shakespeare |
| 1899 | Mich. Term | Northallerton<br>Redcar | } Architecture |
| | | Lincoln<br>Pontefract | } Shakespeare |
| 1900 | Lent Term | York<br>Scarborough<br>Lancaster | Renaissance<br>Novel in the Nineteenth<br>Century |
| 1900 | Mich. Term | Exeter | { Renaissance<br>Shakespeare |
| | | Torquay | { Architecture<br>Shakespeare |
| | | Plymouth | Architecture |
| 1901 | Lent Term | Sunderland<br>Sedbergh<br>Whitby<br>Filey | Architecture<br>Novelists<br>Shakespeare<br>Architecture |
| 1901 | Mich. Term | Colchester<br>Earls Colne<br>Maldon<br>Swaffham<br>Worthing<br>Hastings | } Shakespeare |
| | | Lowestoft<br>Yarmouth | Victorian Poets etc.<br>Nineteenth Century Novel |
| 1902 | Lent Term | Newcastle<br>Sunderland<br>Southport | } Novel |
| | | Middlesbrough<br>Hull | } Architecture |

## TABLE 8 (*cont.*)

| 1902 | Mich. Term | Rawdon | Architecture |
|------|-----------|--------|--------------|
| | | Northallerton | Novel |
| | | W. Hartlepool | |
| | | Hull | Renaissance |
| | | Darlington | Shakespeare |
| 1903 | Lent Term | York | Novel |
| | | Berwick | Shakespeare |
| | | Kelso | |
| | | Sedburgh | |
| | | Hexham | Novel |
| | | Auckland | Architecture |

# Appendix

Amongst the Board's archives are the registers of ten Cambridge tutorial classes held between 1909 and 1926. No other tutorial class registers earlier than 1937 have been traced in any university.* These registers are therefore of great importance for discovering the types of student who attended the first tutorial classes. For five out of the ten classes it has been possible to plot the residences of the students on contemporary street plans of the towns with interesting results.

In all five towns the students tended to be drawn from a few small areas.† If they did not live in the same street they often lived in streets parallel or at right angles to one another. At Nuneaton in 1913 nine students came from the small area between Queen's Road and Coton Road and only four from elsewhere in the urban area.‡ When students were recruited for the second and third years of a class to fill vacancies (as at Rugby in 1918 and 1919 or Bedford in 1920 and 1921) it is noticeable that almost all live near to existing students. When two classes were running simultaneously in a town the pattern of distribution can be very different. At Ipswich in 1919 the Literature class had ten students in the eastern suburbs and eight in the western. The Economic History class, however, had fourteen students in a small area on the western side of the town and the other ten were in two groups north and south of the town centre.

This evidence confirms the reminiscences of former students that they attended because of invitations from friends and acquaintances rather than by seeing circulars or advertisements. Early extension courses, on the other hand, must have depended on widespread publicity – newspaper advertisements, posters and handbills – because of the large number of students attending. Because no early extension course registers have survived it is impossible to make a direct comparison with the class registers. However there is some evidence to show that extension students came from the same areas of the suburbs as class students. The lists of extension students who passed examinations after the early courses at

* Liverpool University Archives have a class register for 1937 and the Oxford Dept. some registers from 1943 onwards. Extension course registers begin at Cambridge in 1960 and at Liverpool and Oxford in the Forties.

† The pattern for Leicester is complicated because some students gave their business rather than their home address.

‡ The remaining students on the register either lived outside the urban area or gave no address.

Leicester show a similar pattern of distribution of students. Moreover, a comparison between modern courses, with their smaller numbers of students, and the early class registers shows remarkable similarities. This has been done for Bedford (one of the only two which are still Cambridge centres) and for a course of 1966 the homes of the students are still to be found on the edges of the urban area. In the north-west of the town they actually coincide with those of 1919.

There is no reason to believe that the Cambridge evidence is untypical of the country at large and unless further evidence can be obtained it must be considered typical. It seems, therefore, that students who attended the early extension courses differed from those attending early tutorial classes only in the method of their recruitment. Neither, despite popular belief to the contrary, has there been any great change in the nature, social standing and motivation of the students over the past sixty years. Any changes which have occurred can be attributed to general population changes in the country, the rise of alternative forms of adult education, and by changes in the popularity of different subjects.

# References

The Board and its predecessor the Syndicate kept a very full record of its activities from the beginning and its early archives have been supplemented by the gift to the Board of the Stuart, Colman and Moulton archives as well as several smaller collections. These archives have been the primary source of information for this volume, and citations to them are given in the notes below under the reference BEMS. At present the archives are still stored at Stuart House, but it is proposed to transfer them to the University Archives as soon as space becomes available for them. Copies of the catalogue can be consulted at the University Archives, the Historical Manuscripts Commission and Stuart House.

Although the Board's archives are so extensive, it should be noted that they reflect very closely the administrative structure of the Syndicate and the Board. Between 1878 and 1924 the minutes of the joint Syndicate are still in the custody of the Local Examinations Syndicate, while the Lecture Committee minutes in the Board's archives only begin in 1892. The archives include copious details about the selection and work of lecturers because that was the Syndicate's responsibility. Very little can be found there about the dates and times of lectures, the fees charged to students and similar details because these were not arranged by the Syndicate but by each local committee or society. For extension courses this information has to be sought in local newspapers.

There are two other serious gaps in the archives which cannot be filled so easily. Almost no information about the examinations organised for extension students can be found either there or with the Local Examinations Syndicate. Neither has more than a handful of registers of students before 1960 survived. A number of records were destroyed some years ago, but it seems more likely that it was never the policy to collect and retain these two classes of records, which had little administrative value.

## CHAPTER 1

1 F. A. Cavenagh, *The Life and Work of Griffith Jones* (Cardiff, 1930), pp. 38, 54, 56.

2 W. K. L. Clarke, *A History of the S.P.C.K.* (London, 1959), p. 25. H. P. Thompson, *Thomas Bray* (London, 1954), pp. 36, 105.

3 J. Telford, ed., *Letters of John Wesley* (London, 1931), vol. VII, p. 258.

4 L. F. Church, *The Early Methodist People* (London, 1948), p. 256.

5 G. C. Martin, *The Adult School Movement* (London, 1924), pp. 12, 29, 73, 81.

6 Sheffield Archives Dept., NR 72.

7 See the *Western Weekly Mercury* for October 1887 for advertisements of their activities.

8 *Trans. Devons. Assoc.*, vol. 94 (1962), pp. 575–8.

9 T. Kelly, *Adult Education* (Liverpool, 1962), p. 113.

10 M. D. Stephens and G. W. Roderick, *The Royal Cornwall Polytechnic Society* (n.d.), p. 4.

11 A. T. Patterson, *Radical Leicester* (Leicester, 1954), pp. 235–8. G. R. Searson, *Liberalism in Leicester* (n.d., c. 1850), p. 28. See also *Rewley House Papers*, vol. III, nos. vii and viii.

12 F. Thompson, *Lark Rise to Candleford* (Oxford, 1954), chap. 31. J. T. Lea, *Mechanics' Institutes* (1968). For an account of the decline of the Institutes see *Chambers' Book of Days* (1864), p. 647.

13 *Derby Mercury*, 14 November 1883.

14 E.g. Devonport Mechanics' Institute (*Western Daily Mercury*, 25 September 1875).

15 Yorkshire. E. Raistrick, *Village Schools* (Yorks. Local History Society, 1971), p. 35.

16 J. Hole, *Literary, Scientific and Mechanics' Institutions* (1853), p. 129. D. Hudson and K. W. Luckhurst, *The Royal Society of Arts* (London, 1954), pp. 244, 245.

17 For the Yorkshire Union see J. F. C. Harrison, *Learning and Living* (London, 1963), chap. 3.

18 Quoted in Kelly, *Adult Education*, p. 119.

19 T. Kelly, *George Birkbeck* (Liverpool, 1957), pp. 72–82.

20 P. Dunsheath and M. Miller, *Convocation in the University of London* (London, 1958), p. 83.

21 *Trans. Devons. Assoc.*, vol. 96, pp. 318–38.

22 Kelly, *Adult Education*, pp. 143, 144. L. Birch, *The History of the T.U.C. 1868–1968* (London, 1968), p. 9.

23 W. Lovett, *Life and Struggles* (London, 1920), vol. II, pp. 253–5. See also E. P. Thompson, *The Making of the English Working Class* (London, 1963), p. 717.

24 R. Owen, *A New View of Society* (London, 1927), pp. viii–x, 99.

25 *Report of Public Schools Commission* (1862), p. 376.

26 A. S. Bishop, *The Rise of a Central Authority for English Education* (Cambridge, 1971), p. 153.

27 *The Leicester Cemetery and the Leicester Town Museum* (1849), p. 17.

28 E.g. Longparish in Hampshire (*Minutes of Committee of the Council on Education, 1846*, vol. I, p. 79).

29 M. E. Sadler, *Continuation Schools in England and Elsewhere* (Manchester, 1907), pp. 56–63.

30 Kelly, *Adult Education*, pp. 182, 183.

31 *Trans. Leics. Arch. Soc.*, vol. 33, pp. 50, 51.

32 J. Rhys and D. Brynmor-Jones, *The Welsh People* (London, 1900), p. 489. Dunsheath and Miller, *Convocation in the University of London*, pp. 1–5.

33 Leicester Univ., Vaughan College, minute book 1862–1907, pp. 74, 181.

34 Dunsheath and Miller, *Convocation in the University of London*, p. 54.

35 Quoted in Kelly, *George Birkbeck*, p. 247.

36 V. Bonham-Carter, *In a Liberal Tradition* (London, 1960), p. 76.

37 Higham, *F. D. Maurice*, p. 56. Leicester Univ., Vaughan College, minute book 1862–1907, p. 261. *West Riding Post Office Directory* 1871.

38 *Victoria County Hist. of Cambs.*, vol. III (1959), p. 260.

39 *Southport and Birkdale University Extension Society 1874–1954* (Leicester, 1954), p. 15. I. C. Ellis, *Nineteenth Century Leicester* (Guernsey, 1935), p. 236. G. Raverat, *Period Piece* (London, 1952), pp. 95, 96.

40 *Prospectus and Rules of the North of England Council . . .* (1868), p. 7.

41 *Report of the First Meeting of the North of England Council . . .* (Manchester, 1868). S. C. Lemoine, 'The North of England Council' (Manchester M.Ed. thesis, 1968), pp. 52, 73, 92. B. Megson and J. Lindsay, *Girton College* (Cambridge, 1960), p. 2.

42 Univ. Archives, C.U.R. 57.1, 42–7.

43 J. Roach, *Public Examinations in England 1850–1900* (Cambridge, 1971), p. 120.

44 The statistics are conveniently summarised in Appendix VIIc of Miss Lemoine's thesis, 'The North of England Council'.

45 *Report of a Conference of Local Secretaries* (Leeds, 1871), pp. 20, 21.

46 F. W. H. Myers, 'Local Lectures for Women' (*Macmillan's Magazine*, December 1868, pp. 159–63).

47 For a history of the College see M. A. Hamilton, *Newnham, An Informal Biography* (London, 1936).

## CHAPTER 2

1 J. Stuart, *Reminiscences* (London, 1911), p. 58. Much of the information in this chapter is taken from this source.

2 Univ. Archives, Subscription Book 1860–9, 402.

3 *Reminiscences*, p. 23.

4 *Reminiscences*, pp. 114–16.

5 *Reminiscences*, pp. 128, 129.

6 *Reminiscences*, p. 137.

7 D. A. Winstanley, *Early Victorian Cambridge* (Cambridge, 1940), pp. 314, 356.

8 D. A. Winstanley, *Later Victorian Cambridge* (Cambridge, 1947), pp. 276–8, 334.

9 R. Watson, *Anecdotes of the Life of the Bishop of Llandaff* (London, 1817), pp. 28, 29.

10 Winstanley, *Later Victorian Cambridge*, p. 192.

11 *Ibid.* pp. 185, 206, 148.

12 *Ibid.* pp. 144–7.

13 Hudson and Luckhurst, *The Royal Society of Arts*, pp. 246–9.

14 Lemoine, 'The North of England Council', p. 35. Roach, *Public Examinations in England 1850–1900*, pp. 66–72.

15 Univ. Archives, C.U.R. 57.1, 1, 2 and 4.

16 Univ. Archives, C.U.R. 57.1, 7, 40, 42 and 47. Roach, *Public Examinations in England 1850–1900*, pp. 109 and 110.

17 *Victoria County Hist. of Cambs.* vol. III (1959), p. 259.

18 Winstanley, *Later Victorian Cambridge*, p. 67. See also D. G. James, *Henry Sidgwick* (Oxford, 1970).

19 *Reminiscences*, p. 152.

20 *Reminiscences*, p. 182.

21 S. Rothblatt, *The Revolution of the Dons* (London, 1968), p. 143. *Reminiscences*, pp. 181, 194.

22 *Reminiscences*, p. 198.

23 Winstanley, *Later Victorian Cambridge*, p. 241.

24 *Ibid.* pp. 243–60.

25 Univ. Archives, Subscription Book 1860–69, 402.

26 *Reminiscences*, pp. 155, 156.

27 T. J. N. Hilken, *Engineering at Cambridge University 1783–1965* (Cambridge, 1967), p. 63.

28 *Reminiscences*, pp. 182–92. Hilken, *Engineering at Cambridge University*, pp. 64–9.

29 Hilken, *Engineering at Cambridge University*, pp. 68, 75–9. BEMS 1/44.

30 Hilken, *Engineering at Cambridge University*, p. 91.

31 BEMS 1/38. *Reminiscences*, p. 175.

32 BEMS 1/24. *Reminiscences*, p. 235.

33 BEMS 1/6, 95–107.

34 *Reminiscences*, pp. 218–21. A. S. G. Butler, *Portrait of Josephine Butler* (London, 1954), pp. 75–96.

35 Hamilton, *Newnham, An Informal Biography*, p. 109. *Reminiscences*, pp. 178–81.

36 H. Burton, *Barbara Bodichon* (London, 1949), p. 174. M. C. Bradbrook, *'That Infidel Place'* (London, 1969), p. 22.

37 In 1874. L. Campbell and W. Garnett, *Life of James Clerk Maxwell* (London, 1882), pp. 631–3.

38 *Reminiscences*, p. 257.
*Reminiscences*, pp. 253–5.
BEMS 1/7.

41 BEMS 43/8 and 1/18.

42 H. C. Colman, *University Extension Lectures in Norwich* (Norwich, 1937), pp. 31, 32.

43 Colman, *Jeremiah James Colman*, p. 130.

44 BEMS 43/8 and 1/19.

45 BEMS 1/20, 19.

46 Butler, *Portrait of Josephine Butler*, p. 63.

47 Trinity Coll. MS. c. 43, 110.

48 G. F. Browne, *Recollections of a Bishop* (London, 1915), p. 226.

## CHAPTER 3

1 *Reminiscences*, pp. 154, 155.

2 BEMS 37/1, 97.

3 *Reminiscences*, pp. 157–64. See also BEMS 1/10.

4 *Inaugural Address at the Second Oxford Summer Meeting*, 1889, p. 22 (BEMS 43/8).

5 BEMS 1/10.

6 *Reminiscences*, p. 163. BEMS 1/8.

7 In 1869 – BEMS 1/9.

8 The letter is reprinted in *Six Lectures to the Workmen of Crewe* (BEMS 1/9).

9 *Inaugural Address at the Second Oxford Summer Meeting*, 1889, p. 27 (BEMS 43/8).

10 BEMS 1/10.

11 *Reminiscences*, p. 166.

12 *University Extension*, p. 5.

13 BEMS 1/2.

14 BEMS 1/6, 1.

15 BEMS 1/3.

16 BEMS 1/6, 1–4.

17 J. L. Paton, *John Brown Paton* (London, 1914), p. 158. See also Curwen MSS. C1/7 and 13 at New College, London.

18 Harrison, *A History of the Working Men's College*, p. 40.

19 *Trans. Leics. Arch. Soc.*, vol. 33, pp. 50–4. Ellis, *Nineteenth Century Leicester*, pp. 175, 236. BEMS 1/2.

20 BEMS 1/6, 3 and 4.

21 *Leicester Chronicle*, 3 May 1873.

22 BEMS 1/25.

23 *Derby and Chesterfield Reporter*, 2 May 1873.

24 *Derby Mercury*, 7 May 1873.

25 BEMS 1/11.

26 Univ. Archives, C.U.R. 57.1, 48 (11 November 1871).

27 Univ. Archives, C.U.R. 57.1, 48 (20 November 1871).

28 Univ. Archives, C.U.R. 57.1, 48 (n.d.).

29 Univ. Archives, C.U.R. 57.1, 48 (n.d.).

30 BEMS 1/3, letter to his mother, 13 March 1872.

31 BEMS 1/3, letter to his mother, 6 April 1872. BEMS 4/1, 3.

32 BEMS 4/1, 5.

33 BEMS 1/5.

34 BEMS 4/1, 6 and 7.

35 BEMS 1/3.

36 BEMS 4/1, 5.

## CHAPTER 4

1 BEMS 4/1, 15.

2 BEMS 4/1, 55.

3 BEMS 4/1, 56.

4 *Derby Mercury*, 17 September 873.

5 S. D. Chapman, *The Early Factory Masters* (Newton Abbot, 1967), pp. 94, 171, 233.

6 *The Derbyshire Red Book, Almanack and Annual Register for 1890*, p. 167.

7 *Derby Mercury*, 1 October 1873.

8 *Derby Mercury*, 17 September 1873.
9 *Derby Mercury*, 24 September 1873.
10 *Derby Mercury*, 8 Ocotber 1873. *Nottingham Journal*, 11 October 1873.
11 *Derby Mercury*, 8 and 15 October 1873.
12 Leicester City Museums, committee minute book, 1871–9, p. 27.
13 *Ibid.* pp. 25, 27, 31.
14 *Leicester Chronicle*, 4 October 1873.
15 *Leicester Chronicle*, 18 October, and *Leicester Journal*, 17 October 1873.
16 *Leicester Chronicle*, 11 October 1873.
17 A copy is in BEMS 33/9.
18 *36th Annual Report of the Institution* (1874), p. 5.
19 BEMS 33/9.
20 BEMS 53/12.
21 *Nottingham Journal*, 11 October 1873.
22 *Nottingham Journal*, 10 and 11 October 1873.
23 E.g. in Paton, *John Brown Paton*, p. 157 and J. C. Warren, *A Biographical Catalogue of Portraits, etc. at High Pavement Chapel* (n.d.), p. 25.
24 *Nottingham Journal*, 10 May 1873.
25 *Report of Eighth Meeting of North of England Council* . . . (1873), p. 9.
26 *Leicester Chronicle*, 3 May 1873.
27 *Nottingham Daily Guardian*, 31 January 1871.
28 *Report on Second Oxford Summer Meeting*, 1889, p. 33.
29 *Bradford Chronicle*, 10 December 1873 (BEMS 1/6, 6).
30 *Halifax Courier*, 13 December 1873 (BEMS 1/6, 7).
31 *Annual Report of the Association*, 1874 (Leeds City Archives, Yorks. Ladies' Council, 21).
32 Leeds City Archives, Yorks. Ladies' Council, vol. 56B, 22–32.
33 *Leeds Mercury*, 6 January 1874. (BEMS 1/6, 10).
34 *Yorkshire Post*, 3 January 1874.
35 *Birmingham Morning News*, 18 December 1873 (BEMS 1/6, 8).

36 BEMS 33/1.
37 *Second Report of Museum Committee*, 1874 (Leicester City Museum 4D56/2/1).
38 Leicester Univ., Vaughan College, minute book 1862–1907, p. 135.
39 *Nottingham Daily Express*, 10 October 1874 (BEMS 1/6, 21).
40 Nottingham Archives Dept., Acc. M 1726 – Accounts for tickets sold 1876–81.
41 Lemoine, 'The North of England Council', p. 100.
42 *Liverpool Mercury*, 21 January 1874 (BEMS 1/6, 13).
43 BEMS 33/7.
44 A. Cunningham, *William Cunningham* (London, 1950), p. 33.
45 BEMS 22/1, 64.
46 BEMS 22/1, 61.
47 BEMS 4/1, 22.
48 BEMS 22/1, 58 and 59.
49 BEMS 4/1, 23.
50 BEMS 37/1, 31, 60. *Report presented to the Syndicate* . . . 1875 (BEMS 53/12).
51 *Report*, pp. 6, 7.
52 *Report*, pp. 7–9.
53 *Report*, pp. 13 and 14.
54 BEMS 37/1.
55 BEMS 37/1, 44.
56 BEMS 37/1.
57 BEMS 37/1, 348.
58 BEMS 37/1, 345, 145.
59 Rothblatt, *The Revolution of the Dons*, p. 233.
60 BEMS 4/1, 29.
61 BEMS 22/1, 67.
62 *Calendar of Local Lectures*, 1880 (BEMS 23/2), pp. 146–9.
63 BEMS 37/1, 335.
64 BEMS 4/1, 34 and 37/1, 206.
65 BEMS 1/37, 3. BEMS 4/1, 24 and 34.
66 BEMS 4/1, 20.
67 BEMS 4/1, 24. BEMS 1/37.
68 BEMS 4/1, 34. BEMS 1/37. BEMS 18/1.

## CHAPTER 5

1 J. Stuart, *University Extension* (Cambridge, 1871).

2 Leeds Archives Dept., Yorks. Ladies' Council archives, vol. 56B, pp. 24, 25.

3 *University Extension Journal*, vol. 5, p. 48. *Calendar of Cambridge Local Lectures* (1880), p. 56.

4 *Calendar of Cambridge Local Lectures* (1880), p. 99.

5 University Archives, C.U.R. 57.1, 84.

6 BEMS 26/1, 56.

7 *Liverpool Mercury*, 21 January 1874 (BEMS 1/6, 13).

8 *Calendar of Cambridge Local Lectures*, 1880, p. 69 (BEMS 33/7).

9 *Ibid.* p. 27.

10 See *University Extension Journal*, vol. 6, pp. 6 and 50.

11 Kelly, *Adult Education in Liverpool*, p. 35.

12 *36th Annual Report* of the Institution, 2 February 1874, p. 5.

13 W. Moore Ede, *Report presented to the Syndicate*, 1875, p. 11. Nottingham Archives Dept., *Council Reports*, 1874–5.

14 Wood, *University College, Nottingham*, p. 16.

15 Nottingham Archives Dept., misc. committee minute book, 1875–81, pp. 4, 10, 26: *Council Reports* 1874–5.

16 Nottingham Archives Dept., misc. committee minute book, 1875–81, pp. 28, 98.

17 BEMS 12/1, ff. 395–417. Enfield's reply is on ff. 450–7.

18 S. Weintraub, *Shaw An Autobiography* (London, 1969), p. 122. Wood, *University College, Nottingham*, p. 38.

19 Local Examinations Syndicate, minute book 2, 17 November 1883.

20 *Calendars of University College Nottingham*, 1882–3 and 1895–6.

21 *Calendar of Cambridge Local Lectures*, 1880, pp. 122–7.

22 BEMS 22/1, f. 72.

23 *Recollections*, p. 48.

24 *Recollections*, pp. 121–4.

25 BEMS 4/1, 43.

26 Local Examinations Syndicate, minute book 2, 3 June and 19 October 1878.

27 *Annual Report*, 1879 – BEMS 22/1, f. 72.

28 University Archives, C.U.R. 57.1, 117.

29 Cunningham, *William Cunningham*, pp. 3, 15, 23, 24.

30 Local Examinations, minute book 2, 7 May and 3 June 1878.

31 *Ibid.* 7 February 1880 and 12 February 1881. Cunningham, *W. Cunningham*, pp. 36–8.

32 B. B. Thomas, 'R. D. Roberts and Adult Education' in B. B. Thomas (ed.), *Harlech Studies* (Cardiff, 1938), p. 1. *Dictionary of Welsh Biography*, p. 878.

33 BEMS 12/1, 59–61.

34 BEMS 12/1, 96 and 97.

35 BEMS 22/1, 75. Local Examinations Syndicate, minute book 2, 5 June 1880 and 12 March 1881.

36 BEMS 22/1, 75. *Annual Report*, June 1882.

37 BEMS 22/1, 76. Local Examination Syndicate, minute book 3, 22 March 1882.

38 BEMS 22/1, 78, 88–91.

39 BEMS 22/1, 88–91.

40 Local Examinations Syndicate, minute book 2, 24 October 1885.

41 Thomas, *Harlech Studies*, p. 4.

42 Cunningham, *W. Cunningham*, p. 36.

43 BEMS 1/6, 13.

44 BEMS 4/1, 39 and 40.

45 BEMS 22/1, 64. See also *University Extension Congress Report*, 1894, p. 62.

46 *Derby Mercury*, 14 October 1874.

47 *Nottingham Daily Express*, 10 October 1874 (BEMS 1/6).

48 Leicester Univ., Vaughan College, minute book, 1862–1907, p. 135.

49 Leicester City Museums, minute book 1871–9, p. 63.

50 Leicester City Museums, minute book 1871–9, pp. 105–55.

51 *Calendar of Local Lectures*, 1880, pp. 40, 52, 66.

52 *Report of a Conference in the Senate House* (1887), p. 70 – BEMS 28/1.

53 See its accounts for 1875 in BEMS 37/1, 47.

54 E.g. 'The Death of Simon Fuge' in *The Grim Smile of the Five Towns*.

55 In the same year the Newcastle upon Tyne Literary and Philosophical Society began to aid Cambridge courses (*University Extension Journal*, vol. 5, p. 35).

56 *Leicester Chronicle*, 7, 14 and 21 October 1882.

57 *Annual Report* of the Society, June 1883, p. 14.

58 *Seventh Annual Report*, 1880 (BEMS 22/1, 95).

59 Dunsheath and Miller, *Convocation in the University of London*, p. 50.

60 H. J. Mackinder and M. E. Sadler, *University Extension* (London, 1890), pp. 75–8. Kelly, *Adult Education*, p. 222.

61 See R. D. Roberts, *University Extension* (London, 1908), p. 16. *Annual Report*, 4 June 1885, p. 1 (BEMS 22/1, 96). Local Examinations Syndicate, minute book 3, 16 November 1886.

62 *Seventh* and *Ninth Annual Reports*, 1879 and 1882 (BEMS 22/1, 95 and 86). Local Examinations Syndicate, minute book 2, 11 May 1882.

63 *22nd Annual Report*, 1895, p. 4.

64 Kelly, *Adult Education*, p. 223.

## CHAPTER 6

1 Local Examinations Syndicate, minute book 3, 14 March 1891.

2 BEMS 5/1, 5.

3 BEMS 12/2. Local Examinations Syndicate, minute book 3, 25 February and 21 May 1892.

4 BEMS 4/2, 154.

5 Bishop, *Central Authority for English Education*, p. 187. Technical Instruction Act, 52 and 53 Vict., c. 76.

6 Kent Record Office C/MC 12/1/1, 22.

7 *Ibid.* 28–62.

8 BEMS 22/2.

9 Kent Record Office, C/MC 12/1/1, 112. Local Examinations Syndicate, minute book 3, 14 November 1891.

10 Kent Record Office, C/MC 12/1/1, 213.

11 Devon Record Office, Technical Education minute book, 1890–1900, 1–9.

12 Devon Record Office, Technical Education minute book, 1890–1900, 13–15.

13 BEMS 22/2.

14 *County Council Times*, 28 April 1893, p. 334 (BEMS 38/7).

15 *Report on Instruction given in Technical Education* (1891), pp. 3, 4 (BEMS 22/2).

16 *Second Report on Instruction* . . . (1892), pp. 8, 9 (BEMS 22/2).

17 BEMS 38/11.

18 *County Council Times*, 28 April 1893. It was repeated later (BEMS 5/1, 34).

19 Cambs. Record Office, County Technical Education minute book, 1893.

20 Cambs. Record Office, County minute book – *Report* of July 1894.

21 *Third Report on Instruction* . . . (1893), p. 15 (BEMS 22/2).

22 *21st Annual Report*, p. 2.

23 Kent Record Office, C/MC 12/1/1, 66.

24 *Cambridge Independent Press*, 28 September 1894.

25 BEMS 38/12.

26 *University Extension Journal*, vol. 3, p. 51. *University Extension, Exeter Centre Report*, 1890, p. 6.

27 BEMS 5/1, 4.

28 Public Record Office, Ed. 31/7.

29 *Visit of the Chancellor of the University of Cambridge*, n.d. (BEMS 31/2). See the undated report about the advantages of the experiment in BEMS 38/12.

30 Sir Hector Hetherington, *The University College at Exeter 1920–1925* (Exeter, 1963).

31 Hetherington, *University College at Exeter*, p. 17. *Report of the Universities Extra-Mural Consultative Committee*, 1925–6.

32 Public Record Office, Ed. 29/43. BEMS 33/2.

33 Colchester Univ. Ext. Soc., account book, 1919–55 and minute book, 1929–14.

34 *University Extension Journal*, vol. 1, p. 66.

35 Plymouth Archives Dept., council minutes 1910–11, pp. 607–1504. *University Extension Bulletin*, summer 1911, p. 12. Cranage, *Not Only a Dean*, p. 100.

36 BEMS 37/8.

37 BEMS 37/8.

38 *Report of a Conference*...*1890* (1891), pp. 22, 23 – BEMS 28/2.

39 Local Examinations Syndicate, minute book 3, 12 October 1889.

40 *Report of a Conference*...*1890* (1891), pp. 21–32 – BEMS 28/2.

41 Mackinder and Sadler, *University Extension*, p. 134.

42 Mackinder and Sadler, *University Extension*, pp. 139–41.

43 *Manchester Guardian*, 22 November 1893 (BEMS 1/7, 31).

44 *University Extension Journal*, vol. 1, p. 6 and vol. 2, p. 98. BEMS 5/1, 48.

45 BEMS 5/1, 19 and 5/2, 12. *21st Annual Report*, 1894, p. 10.

46 *16th Annual Report*, 1889, pp. 21, 22. BEMS 1/7, 7 and 38/9.

47 BEMS 47/1 – 5. *Report of a Conference*...1898, p. 40 – BEMS 28/3. The accounts for Pioneer Lectures can be found in BEMS 38/13.

48 BEMS 47/9 and 10. R. D. Roberts, *Eighteen Years of University Extension* (Cambridge, 1891), pp. 69–73. BEMS 38/9.

49 *University Extension Journal*, vol. 3, p. 90.

50 Local Examinations Syndicate, minute book 4, p. 215 – letter of Miss Kennedy to Roberts, 3 March 1898. BEMS 38/9 – Regulations, 12 July 1898.

51 Aberystwyth, 1887 – BEMS 22/1, 32.

52 Cambridge, 1891 – BEMS 2/26.

53 Cambridge, 1908 – BEMS 2/29.

54 See BEMS 53/12 and Thomas, *Harlech Studies*, p. 15.

55 See Dunsheath and Miller, *Convocation in the University of London*, chap. 5.

56 Local Examinations Syndicate, minute book 5, f. 8.

57 Thomas, *Harlech Studies*, pp. 22–5.

58 In 1902 he had been appointed High Sheriff of Cardiganshire (*University Extension Journal*, vol. 7, p. 99).

59 Afterwards Assistant Secretary to the Cabinet and first Secretary of the Pilgrim Trust.

60 Then assistant secretary of the London Society and later Vere Harmsworth Professor of Naval History at Cambridge.

61 Local Examinations Syndicate, minute book 5, f. 13.

62 *Ibid.* ff. 16–18.

63 *Ibid.* f. 19.

64 Roach, *Local Examinations in England*, p. 153.

65 Local Examinations Syndicate, minute book 5, f. 20. See Cranage, *Not Only a Dean*, p. 78 for an account of the interview.

## CHAPTER 7

1 Cranage, *Not Only a Dean*, pp. 4–6.

2 BEMS 49/7 and 11. *Annual Report of King's College Council*, 1957–8, p. 23.

3 Cranage, *Not Only a Dean*, p. 8.

4 *Wellington Journal*, 21 November 1891.

5 W. Ward, *Brotherhood and Democracy* (n.d.), p. 129.

6 Cranage, *Not Only a Dean*, p. 8. BEMS 49/10.

7 BEMS 26/1 and 44/8.

8 Cranage, *Not Only a Dean*, pp. 13, 45. BEMS 49/12.

9 A humorous account of one of his lectures can be found in *University Extension Bulletin*, Lent 1909, p. 16.

10 Cranage, *Not Only a Dean*, pp. 46, 47, 218. B.B.C. Written Archives Centre, programme records.

11 Cranage, *Not Only a Dean*, pp. 214, 215. *Annual Report of King's College Council*, 1957–8, p. 25.

12 BEMS 55/10.

13 In *Cambridge Essays on Adult Education*, ed. R. St. J. Parry (Cambridge, 1920).

14 Cranage, *Not Only a Dean*, p. 217.

15 Act, 2 Edw. VII, c. 42.

16 Vol. 8, pp. 66 *et seq.*

17 BEMS 34/7, f. 23. Derby Univ. Ext. Soc., minute book 1906–66, f. 17.

18 A. Mansbridge, *An Adventure in Working Class Education* (London, 1920), pp. 10, 11.

19 'The Veneer' (January 1903), 'A Plan of Action' (March 1903) and 'An Association' (May 1903).

20 Kelly, *Adult Education*, p. 251. Mansbridge, *Working Class Education*, p. 44.

21 BEMS 30/2 and 3 and 45/1. Mansbridge, *Working Class Education*, p. 21.

22 Local Examinations Syndicate, minute book 5, ff. 47 and 59.

23 *Ibid.* ff. 78, 99, 132.

24 Churchill College Archives, ESHR 19/2, p. 5. Quoted by permission of Lord Esher.

25 Derby W.E.A. branch, first minutes, 27 September 1905. Derby Univ. Ext. Soc., minute book, 1888–1906, f. 50, and minute book, 1906–66, f. 14.

26 Local Examinations Syndicate, minute book 5, ff. 170, 179 and 180.

27 Based on BEMS 24/5, *Ipswich Directory* for 1920 and the Poor Rate Books for 1919–20 (East Suffolk Record Office).

28 *Report on Tutorial Classes*, 1912 – BEMS 30/2.

29 BEMS 24/1–3.

30 *Special Report on Certain Tutorial Classes in connection with the Workers' Educational Association* (n.d.) – BEMS 30/1.

31 E.g. the Greens in Flora Thompson's *Lark Rise to Candleford* (chap. 29).

32 *Kettering Leader*, 20 September 1912, p. 6.

33 Flysheet of G. Grant Morris, 22 February 1932 – BEMS 38/22.

34 BEMS 30/1.

35 BEMS 30/3.

36 *Tutorial Classes Committee Report*, 1914. *Cambridge Daily News*, 3 May 1911.

37 BEMS 6/1, 1 and 2.

38 *Oxford and Working Class Education* (1908), p. 55.

39 BEMS 6/1, 12–24.

40 BEMS 52/3 and 4.

41 BEMS 6/1, 50–67.

42 BEMS 6/1, 83.

43 J. Stuart, *An Inaugural Address ... at the ... Third Series of Vacation Courses* (1892), p. 12 – BEMS 1/19.

44 Roberts, *Eighteen Years of University Extension*, p. 86.

45 *Annual Report*, 29 May 1886 – BEMS 22/1, f. 100.

46 *Annual Report*, 1 May 1888 – BEMS 22/1, f. 106. Roberts, *Eighteen Years of University Extension*, p. 87.

47 *Report of the Syndicate on the Course of Study at Cambridge* (1891), p. 1. See also *Report of a Conference ... 1890* (1891), pp. 32, 33.

48 *Ibid.* p. 15.

49 BEMS 1/19.

50 *20th Annual Report*, 1893, p. 3.

51 BEMS 5/1, 7 and 11. *20th Annual Report*, 1893, p. 4.

52 Local Examinations Syndicate, minute book 4, p. 22.

53 *23rd Annual Report*, 1896, p. 4.

54 BEMS 5/1, 17. *24th Annual Report*, 1897, p. 4.

55 BEMS 5/3, 7 and 22. Cranage, *Not Only a Dean*, pp. 86, 99. Originally Durham was suggested in 1904.

56 *The British Empire*, Cambridge Summer Meeting (1912), p. 102 – BEMS 15/10. *Kindness, Beauty and Learning* (1906), p. 5 – BEMS 15/7. There is an analysis of students attending summer meetings in *University Extension Journal*, vol. 1, p. 3 and vol. 2, p. 22.

57 BEMS 5/3, 120 and 131.

58 BEMS 15/4.

59 *University Extension Bulletin*, Summer 1910, p. 13.

60 *University Extension Bulletin*, Michaelmas 1910, p. 9.

61 R. Rössger, *Die University Extension-Bewegung* (Leipzig, 1904), p. 47.

62 BEMS 5/4, 126. Cranage, *Not Only a Dean*, p. 106.

63 BEMS 5/4, 128.

64 BEMS 5/4, 131 and 162. Cranage, *Not Only a Dean*, pp. 107, 108.

REFERENCES (pp. 118–131)

65 Cranage, *Not Only a Dean*, p. 114.
D. H. S. Cranage (ed.), *The War and
Unity* (Cambridge, 1919).
66 Cranage, *Not Only a Dean*, p. 111.
Local Centres Union, E. Anglian
minute book, 1914–40, f. 6.
67 BEMS 30/5 and 24/4–8.

68 *Thomas Cecil Fitzpatrick. A Memoir* (Cambridge, 1937), pp. 36–41.
69 *University of Cambridge Jubilee
Celebration of the Local Lectures*
(1923), p. 3.
70 *Ibid.* p. 23. By Prof. R. Peers of
Nottingham.

CHAPTER 8

1 W. F. Moulton, *Richard Green
Moulton* (London, 1926), p. 24. E.
Carpenter, *My Days and Dreams*
(London, 1916), p. 80. *Cambridge
Bulletin*, no. 1, p. 6.
2 University Archives, C.U.R. 57.1,
57.
3 BEMS 37/1.
4 *University Reporter*, 1891, pp. 195,
196.
5 BEMS 55/24 – Powys.
6 BEMS 55/24 – Priestley. BEMS
55/12 – Eliot.
7 BEMS 55/9 – Bruce. BEMS 55/12
– Eliot.
8 BEMS 38/12.
9 BEMS 55/8.
10 BEMS 38/9.
11 BEMS 55/10.
12 BEMS 55/25 and 55/8.
13 BEMS 55/10.
14 Interview with Hardman, 1970.
15 Cranage, *Not Only a Dean*, pp. 45,
46.
16 BEMS 55/10.
17 BEMS 55/24.
18 BEMS 55/12.
19 BEMS 55/11.
20 BEMS 55/21.
21 BEMS 5/3, 11. He was later in
dispute with the Syndicate about this.
22 R. C. Fitzpatrick to Cranage, 15
October 1908 (BEMS 55/11).
23 BEMS 37/1, 37.
24 *21st Annual Report*, 1894, p. 4.
25 BEMS 38/9.
26 BEMS 55/6. *45th Annual Report*,
1918.
27 *28th Annual Report*, 1901.
28 Moulton, *Richard Green Moulton,
A Memoir*, pp. 11–15.
29 *Ibid.* pp. 34, 35.
30 BEMS 37/3, 54.
31 BEMS 37/3, 69.
32 BEMS 37/3, 70.

33 Moulton, *R. G. Moulton*, p. 74.
34 *Ibid.* pp. 79–101.
35 *Ibid.* pp. 103, 114.
36 BEMS 3/91.
37 Moulton, *R. G. Moulton*, pp. 124,
125.
38 BEMS 55/27.
39 BEMS 55/27 – letter of 3 March
1903. See Table 8.
40 Exeter, Plymouth and Torquay on
'Chaucer's England' (BEMS 26/1).
41 On 'Dante's Italy' (BEMS 26/1).
42 G. G. Coulton, *Fourscore Years*
(Cambridge, 1943), p. 283. Quoted by
kind permission of his daughter.
43 *Ibid.* p. 284.
44 1892. Moulton, *R. G. Moulton*, p.
103.
45 BEMS 36/7, 12.
46 BEMS 36/7, 94–100.
47 BEMS 24.
48 R. G. Moulton, *The University
Extension Movement* (1885), pp. 15–
21.
49 *Derby Mercury*, 17 September
1873.
50 *Calendar of Local Lectures, 1875–80*
(1880), pp. 76, 103.
51 *Cambridge Independent Press*, 17
September 1887.
52 Exeter University Registry, Exeter
Univ. Ext. Committee, minute book,
1886–88, ff. 1, 2.
53 *Calendar of Local Lectures, 1875–80*
(1880), pp. 103, 45, 84. *University
Extension Journal*, vol. 3, p. 22.
54 Reports for 1883/4, 1889/90,
1890/1 are in Derby Public Library,
and for 1886/7 in BEMS 22/1, f. 19.
55 *Derby Mercury*, 14 November
1883.
56 *University Extension Journal*, vol.
1, pp. 131–3 and vol. 4, pp. 5–7.
*University Extension Societies* (n.d.
c. 1898) – BEMS 53/12.

57 *Ibid.* f. 6. Most of them refused.
58 BEMS 22/1, ff. 88–93.
59 *Annual Report* of the Derby Society, September 1886.
60 BEMS 22/1, 53. BEMS 34/3.
61 BEMS 34/4. See above, p. 88.
62 BEMS 34/1, 2 and 5. BEMS 12/2, 134 and 164.
63 *22nd Annual Report*, 1895 – BEMS 35/2.
64 BEMS 5/2, 117. BEMS 53/3.
65 BEMS 17/5.
66 *Western Daily Mercury*, 30 September 1875.
67 Cranage, *Not Only a Dean*, p. 45.
68 BEMS 22/1, ff. 89, 109, 110.
69 Sheffield Archives Dept., Carpenter Papers C374.6S.
70 Hastings Public Library, Browning MSS., letter of A. Berry, c. 1892.

71 *Ibid.* letters of Browne, 4 May 1885 – 8 April 1890.
72 BEMS 37/3, 23.
73 University Archives, C.U.R. 57.1, 83–8. See above, p. 72.
74 BEMS 22/1, f. 107. BEMS 35/8.
75 University Archives, C.U.R. 57.1, 142.
76 *4th Annual Report of Board*, 1928, p. 4.
77 *36th* and *14th Annual Reports of Board*.
78 *Report Presented to the Syndicate* (1875), p. 12.
79 BEMS 38/14.
80 A copy heavily amended by Roberts is in BEMS 38/11.
81 *University Extension Journal*, July 1898, pp. 136, 137. *The Times*, 26 May 1898. BEMS 38/14.
82 Roberts, *University Extension*, p. 22.

## CHAPTER 9

1 *44th Annual Report*, 1917, pp. 2–4. Liverpool Univ. Archives, Southport Univ. Ext. Soc., meeting minute book, 1896–1943, f. 30.
2 *Final Report of the Adult Education Committee of the Ministry of Reconstruction* (1919), p. 1.
3 *Final Report*, pp. 78, 82.
4 *Final Report*, pp. 83–6.
5 *Final Report*, p. 87.
6 *Final Report*, chap. 12.
7 *Final Report*, pp. 97, 98.
8 See S. G. Raybould, *University Extramural Education in England* (London, 1964).
9 *Final Report*, p. 170.
10 *Final Report*, p. 103.
11 *Final Report*, pp. 169, 170.
12 *University Reporter*, 1919–20, pp. 56–8.
13 Local Examinations Syndicate, minute book 4, p. 194.
14 *University Reporter*, 1923–4, p. 892.
15 BEMS 4/2, 2.
16 Cranage, *Not Only a Dean*, p. 130.
17 BEMS 4/2, 9, 10.
18 BEMS 4/2, 4. BEMS 39/3. Information from Mr B. W. C. Green.
19 BEMS 39/3. *Stuart House, Cambridge* (1927) – BEMS 39/5.
20 Cranage, *Not Only a Dean*, p. 98.

21 BEMS 39/3.
22 BEMS 39/1.
23 BEMS 4/2, 52. Information provided by Mr Hickson and Mr Green.
24 BEMS 39/2 and 3. Information provided by Mr Hickson.
25 BEMS 39/5.
26 *The Times*, 29 Jan. 1927, p. 13.
27 Cranage, *Not Only a Dean*, p. 159.
28 BEMS 4/2, 116.
29 Kelly, *Adult Education*, p. 271.
30 BEMS 37/19. *Southport and Birkdale Extension Society, 1874–1954* (1954), p. 13. Wellingborough Univ. Ext. Soc. records, 1962–72.
31 BEMS 37/20.
32 BEMS 37/21.
33 BEMS 26/1.
34 BEMS 22/1, f. 101. Local Examinations Syndicate, minute book 4, 14 November 1891.
35 Local Centres Union, E. Anglian minute book, 1914–40, ff. 9, 11, 21. BEMS 6/1, 132.
36 See Kelly, *Adult Education*, p. 271 and *The Development of Adult Education in Rural Areas* (1922).
37 BEMS 4/2, 98.
38 BEMS 10/1, 18. BEMS 4/2, 174.
39 BEMS 10/1, 48, 82, 94. BEMS 4/2, 204, 305. *Annual Report*, 1938, p. 3.

40 BEMS 29/1.
41 BEMS 29/2.
42 BEMS 38/16 – *Grants in Recent Years.* See also Cranage, *Not Only a Dean*, p. 81.
43 BEMS 38/19.
44 Act 11 and 12 Geo. V, c. 51. Grant Regulations no. 33 of 1924. See Parry (ed.), *Cambridge Essays on Adult Education*, p. 175.
45 Grant Regulations no. 14 of 1938.
46 BEMS 48/5. *Cambridge Bulletin*, no. 1, pp. 18, 19.
47 BEMS 36/7, pp. 133, 139, 153. Information from Mr David Hardman.
48 BEMS 6/1, 151.
49 BEMS 7/1, 128. BEMS 48/9.
50 BEMS 48/10 – letter of 25 October 1937.
51 BEMS 48/11 – W.E.A. memorandum, n.d., para 13.
52 BEMS 48/11 – report of subcommittee, n.d. BEMS 6/2, 124.
53 BEMS 6/2, 160.
54 BEMS 6/2, 170.
55 BEMS 4/2, 298. BEMS 48/11.
56 *Cambridge Bulletin*, no. 2, p. 37.
57 BEMS 33/22, f. 48.

58 BEMS 4/2, 64.
59 *New Ventures in Broadcasting* (B.B.C., 1928), p. 87.
60 *Cambridge Bulletin*, no. 4, p. 90.
61 *Cambridge Bulletin*, no. 5, p. 27.
62 *Cambridge Bulletin*, no. 7, p. 6.
63 Local Centres Union, E. Anglian minute book, 1914–40, f. 45.
64 BEMS 4/2, 284. BEMS 36/7, 155. For the comparative failure of the B.B.C.'s work in this field see Kelly, *Adult Education*, pp. 318–20.
65 *Report of Royal Commission on Oxford and Cambridge Universities* (1922), pp. 180–2.
66 BEMS 56/8 – Scheme of Bursaries, p. 1. BEMS 8/1, 80.
67 BEMS 56/8 – Scheme of Bursaries, p. 9.
68 *Ibid.* p. 8.
69 BEMS 8/1, 34.
70 BEMS 8/1, 20.
71 *University Reporter*, 1927/8, pp. 1027, 1046.
72 BEMS 48/4. See Table 7.
73 BEMS 34/7, f. 63.
74 E. G. Harwood (1925–6) in *Cambridge Bulletin*, no. 1, p. 3.

## CHAPTER 10

1 BEMS 10/1, 12.
2 *15th Annual Report*, 1939, p. 3.
3 *16th Annual Report*, 1940. See also Local Centres Union, E. Anglian minute book, 1914–40, ff. 58, 59.
4 Wellingborough Univ. Ext. Soc., secretary's report for 1939.
5 BEMS, Ely University Extension Committee minute book, 1938–57, ff. 9–11.
6 BEMS 4/2, 374.
7 Derby University Extension Society, *Annual Report*, 1939–41.
8 BEMS 2/14. See also BEMS 4/2, 392.
9 *21st Annual Report*, 1945. Information from Mr David Hardman.
10 BEMS 6/2, 182.
11 *The Times*, 9 February 1940.
12 BEMS 57/3.
13 BEMS 57/1. Afterwards Mr Wilson wrote *Education in the Forces, 1939–46: The Civilian Contribution* (1949).

14 BEMS 4/2, 390.
15 BEMS 57/1.
16 BEMS 48/15.
17 BEMS 48/17.
18 BEMS 48/20 and 21. *The Times Educational Supplement*, 16 August 1947.
19 BEMS 38/25.
20 BEMS 38/25 and 48/17.
21 BEMS 48/19.
22 Kelly, *Adult Education*, pp. 373, 374.
23 BEMS 14/21.
24 *20th Annual Report*, 1944, p. 3. BEMS 16/31 and 32. BEMS 14/22.
25 *24th Annual Report*, 1948, p. 4.
26 Kelly, *Adult Education*, p. 385.
27 BEMS 48/19.
28 *University Reporter*, 1948–9, pp. 385 and 1618.
29 BEMS 48/19.
30 BEMS 4/3, 10. *Report of Universities' Council for Adult Education*, 1945–7, p. 7.

## REFERENCES (pp. 166–177)

31 *27th Annual Report*, 1951, p. 3.
32 E.g. Wellingborough University Extension Society, annual report, 1960.
33 *28th–39th Annual Reports.*
34 *41st and 43rd Annual Reports.*
35 BEMS 14/4.
36 *50th Annual Report of Syndicate*, p. 4.
37 *Cambridge Bulletin*, no. 3, p. 53.
38 *5th Annual Report*, p. 4.
39 BEMS 50/43 and 48/11.
40 *40th Annual Report*, 1964.
41 BEMS 8/2, 6. BEMS 48/20.
42 BEMS 8/2, 14, 36, 72, 110.

43 *27th Annual Report*, 1951, p. 11.
44 BEMS 4/3, 46. *30th Annual Report*, 1954.
45 *The Organisation and Finance of Adult Education in England and Wales* (1954).
46 *Crisis in Adult Education* (1964).
47 Board circular of 27 May 1971.
48 *35th and 46th Annual Reports.*
49 *38th Annual Report.* See *Calendar of Local Lectures*, 1880, p. 56.
50 *41st Annual Report.*
51 L. Munby (ed.), *East Anglian Studies* (Cambridge, 1968).

# Index

Chicago University, R. G. Moulton at, 102, 125
Christian
Institutes, 2
Socialism, 7, 72
Church
unity of, talk on, 118
Congress at Cambridge, 116
Churches, Central Council for, 146
*Churches of Shropshire*, by D. H. S. Cranage, 105
Churchill, Sir Winston, and adult education, 173
cinema and adult education, 156
Circulating Schools in Wales, 1, 77
citizenship, courses on, 107
civil servants, 107, 159, 190
Clapham, J. H., lecturer, 127
Clark
Miss, Royston Centre, 117
Rev. Walter, Derby Centre, 34, 47, 48
Clarke, E. T., lecturer, 139
classes, 176
Local Lecture, 30, 46, 134, 135;
statistics, 178–81
for women, 7
Clayden, A. W., Exeter University College, 95, 96
clerks, students, 112, 128, 133, 190
Clough, Miss A. J., 9, 11, 26
and Miss Davies, 10, 56
and J. Stuart, 25, 26
Cobham, Alfred, Southport Centre, 119
Colchester (Essex)
College, 95, 96
Corporation, 96, 172
Extension Society, ix, 96, 116, 132, 148, 172, 191
Public Library, ix
Coleg Harlech, 166
colleges, local
affiliated to Cambridge, 138
proposed, 143
Colman, family archives, 195
Colman, Miss Helen C., 144, 146
Colman, J. J., 23
Colman, Laura E., married J. Stuart, 23
Committee for University Assistance, 166
conferences organised by Syndicate, 133

Congregational Church
and adult education, 6
and J. Stuart, 14
Institute, Nottingham, 32
conscientious objectors not employed, 164
conscription and adult education, 141, 165, 166
Conservative Party and adult education, 47, 99, 173
constitutional history, courses on, 54, 60
Contagious Diseases Act and J. Stuart, 21
contumacy, penalty of, 105
Convocation and D. H. S. Cranage, 105
cookery classes, 8, 104
Cooper, Thomas and adult education, 5
co-operation in adult education, 161
*Co-operation, Trade Unionism and Extension*, 107
Co-operative Congresses, 32, 107
*Co-operative News*, article in, 32
Co-operative
Societies and adult education, 5, 30, 52, 59, 116, 132
Wholesale Society, 107
co-partnership, talk on, 117
Cornwall
courses in, 95
Royal Polytechnic Society, 3
Cottenham (Cambs.) Village College, 176
Coulton, G. G., lecturer, 105, 126
country houses and adult education, 167
County Councils, grants from, 87–94
courses, 134, 176
subjects of, 176, 189
Coventry (Warws.) Centre, 125
craft courses, 142, 164
Cranage, D. H. Somerset, Secretary, 114, 117, 118, 120, 122, 128, 132, 143, 155, 177
early life, 104, 105
lecturer, 104, 124, 136
Syndicate Secretary, 102, 103, 104, 106
Board Secretary, 145
Gilchrist Trust Secretary, 99
visited U.S.A., 146, 169
resigned, 146, 147
Dean of Norwich, 146